LESLIE PELTIER'S
Guide to the Stars

LESLIE PELTIER'S
Guide to the Stars

By Leslie C. Peltier
with illustrations by the author

AstroMedia
Milwaukee

Cambridge University Press
Cambridge   London   New York   New Rochelle
Melbourne   Sydney

Published by AstroMedia, a division of Kalmbach Publishing Co.
Milwaukee, Wisconsin 53233
and by
The Press Syndicate of the University of Cambridge
The Pitt Building, Trumpington Street, Cambridge CB2 1RP
32 East 57th Street, New York, New York 10022, USA
10 Stamford Road, Oakleigh, Melbourne 3166, Australia

© 1986 Gordon Peltier. All rights reserved.
Printed in the United States of America.
90 89 88 87 86 5 4 3 2 1

Cover photographs by Jean Dragesco (the Moon) and Alfred Lilge (the Sagittarius starfield).

Library of Congress Cataloging-in-Publication Data

Peltier, Leslie C., 1900–1980
    Leslie Peltier's guide to the stars.

    Includes index.
    1. Astronomy — Amateurs' manuals. 2. Astronomy — Observers' manuals. I. Title. II. Title: Guide to the stars.
QB63.P43 1986   523   86-15183
ISBN 0-913135-04-6
ISBN 0-521-33595-7 (Cambridge)

Contents

	Foreword	vii
1	Trail's Beginning	1
2	The Little Dipper and Polaris	8
3	Augmented Eyes	13
4	Maps of the Sky	19
5	The Winter Stars	21
6	The Spring Stars	36
7	The Summer Stars	50
8	The Autumn Stars	65
9	The Milky Way	79
10	Variable Stars	86
11	Novae and Nova Hunting	95
12	The Sun	100
13	The Planets	105
14	The Moon	124
15	Comets	141
16	Meteors	149
17	Endless Trails	156
	Appendices	
I	Recommended Reading	159
II	The Skies from a Swivel Chair	161
III	A Pronounciation Dictionary	163
IV	The Brightest Stars	165
V	The Constellations	167
VI	Life List of Celestial Objects	171
	Index	183

Foreword

By his life and by his writings, Leslie C. Peltier inspired a whole generation of people to look toward the stars, to learn them and enjoy them. I was part of that new generation, moved by the words of that simple stargazer from Ohio, who through years of observing had become as comfortable with the sky as most folks are with the rooms of their homes.

Leslie Peltier's reputation is based on three things. Of prime importance is his record of discovery; between 1925 and 1954 he found twelve new comets and four exploding stars, or novae.

Less spectacular but every bit as important is Peltier's careful measuring of variable stars over a span of sixty-two years. The 132,000 observations he made provide a record of how these strange stars change their brightness over days, months, and even years. Because he kept watch so carefully, our understanding of these stars is better.

A third area is his writings. (This book is his fourth.) They represent his most significant contribution to astronomy because they don't just point out the stars, they point out a way of enjoying them.

It was during my second visit to Leslie's house that I realized how much Peltier's books and telescopes were a part of him. He took me into a corner of his observatory, and there sat the old mahogany tube that in an earlier time had held the lenses of his 6-inch comet-seeker telescope. There, written in deep, proud letters that almost encircled the little tube, were the designations of the twelve Peltier comets whose light had passed through that telescope before any other. In hand-cut numbers and letters was the core of an inspiring career.

Fifteen times this little telescope had uncovered a new member of our solar system. And the unassuming person standing next to it was responsible for twelve of them.

What Peltier did could profitably be done by amateurs today; what he achieved is a worthwhile goal to which any modern stargazer can aspire. And the mood, the attitudes, the way of thinking he brought to amateur astronomy have given us a special pride in what we do. He is missed by all his friends from several generations, and as you read the pages of this book, I know that he will become your friend too.

—*David Levy*

O ur big Packard twisted into the driveway of just another Midwestern farm house, squatting like all of them on the glacial-drifted Ohio farmland. A couple of cows grazed in the yard, a Model T sagged behind the house, and a board gave passage over a muddy spot in the yard.

The plank-assisted path led into the nearby pasture, and in that pasture was a marvelous sight — a ten-by-ten-foot frame shed with a dome on top. We had reached, for us at any rate, one of the great shrines of the world — the amateur observatory of Leslie C. Peltier. And it was sort of symbolic that the town nearby was called Delphos, the ancient source of revelation.

A slim, quiet man, hardly older than I, greeted us. A vibrant sister Dorothy made the smiling welcome, and, as was customary in the Midwest, we entered through the kitchen door. None of us had seen the other before but we embraced a mutual hobby — we observed variable stars.

This was my introduction to Leslie Peltier, who at the age of 32 was established as America's leading amateur astronomer. Eventually he was to become even more luminous than he was that spring day in 1932. But already he had racked up two comets or so, three novae, and more variable star observations than any one else in the country.

Not long before he died in 1980, we talked of Halley's Comet, which he had seen in 1910. He desperately wanted to see it in 1985, and he knew he would not. Thus, he told me, he was writing a legacy for the next generation. The coming return of Halley, he pointed out, would sell thousands of telescopes to people who otherwise would never have bought them. What attic would embalm them once the comet was long gone?

He told me of a book that might keep many of the telescopes active, available to the young as well as the old. It would be disaster to turn them over to the spiders in the attic. *Leslie Peltier's Guide to the Stars* is that book.

The country supper food was good but I do not remember a dish. It was clear outside, and we would be able to observe that night and match our newly acquired observational skills against the Master

Leslie C. Peltier

of them all. We would take turns estimating the exact brightness of perhaps four dozen variable stars.

If you do not know about them, variable stars change their brightness with time. The variation is periodic for some, ranging from less than a day to over 1,000 days. Some nights binoculars will show the variable; other nights it might take a 20-inch telescope to see the star at minimum light. You can chart these stars and measure the brightness of the stars around them to make a comparison scale. Then on any night you simply compare the variable with the comparison scale and estimate its brightness.

Amateur astronomers since the 1860s have found variables fascinating, and because many of them were within reach of even 3-inch telescopes, the hobby was affordable. By the 1880s magazines were printing these observations. Professor Pickering at Harvard encouraged this amateur work, and he even let some money slip out to help them. He saw it as a way to collect needed data that professional astronomers simply did not have the time to gather.

By 1911 so many amateurs were sending results to Pickering that the American Association of Variable Star Observers was formed. It is now simply called the AAVSO and makes a powerful symbiosis with the professional astronomical machine. Today satellite observing schedules are determined from data collected by AAVSO members.

Peltier joined the AAVSO in 1918, so on that night in 1932 he had fourteen years of experience to my one year. We trudged out that night under skies that Delphos seldom has today. Up went a ladder, up went Peltier, and down was handed the dome slit cover. Farm boy as he was, there was no need for a fancy mechanism to take a cover off.

Inside sat a telescope, gleaming wood, that was already notable. The whim of a Princeton astronomer to loan him the scope shoved Peltier onto the fast track to astronomical fame, a position he maintained with ease and distinguished style for the next half century.

At Princeton it had captured three comets; Peltier had added three more. The names were deeply carved on its mahogany tube. It was a 6-inch, which was large for the AAVSO at that time. As we admired it Peltier noted that if it had been a non-comet telescope he probably would never have gone on to find his but would have stuck only to variables.

The observing session started as soon as it was really dark. We took turns estimating the brightnesses of variable after variable. Leslie found all the fields to speed things up. His scope had no finder; he merely looked along the tube, nudged the scope a trifle, then lo and behold, the variable field would be in the eyepiece. He never was wrong! We had heard of this talent of his, but still it dazzled us.

Later after a ham and egg breakfast, and coffee with grounds, we compared the results of the observing session. I was no more than one-tenth of a magnitude off Peltier. Talk of mead and honey! To me it meant I had graduated, I had gotten around the last buoy in the race, my doubts about myself were quenched by a Peltier fire hose. A great adventure could now begin.

But while I was instantly grateful to Peltier, I did not realize that I was to be only one of many whom Leslie conducted quietly into what may be called significant activity in astronomy, activity that was a contribution to the world I played in.

I didn't dream that Leslie would send dozens of young people into professional astronomy. Some of them now man giant telescopes. More have found that serenity that marked Peltier and have learned that the stars are better than a witch doctor. First it was his example, his role model, then it was his book, *Starlight Nights*, which had the widest impact of any astronomy book since Garret P. Serviss wrote at the turn of the century. Perhaps time will show that Peltier did even more good.

Canny, shy, humble Leslie had special talents unteachable in school. A photographic memory let him remember accurately a star chart even with its comparison star magnitudes over a twenty-year span. He knew the names of all the craters on the Moon even though he made no attempt to memorize them.

Steeped in the 19th-century traditions of the naturalists, Leslie knew the names of all the birds and flowers. His notes record the aurora, the lightning strikes, and the first time each spring the Houstonia bloomed. As a grade school child he carried a cyanide jar through the fields to collect butterflies. The culture of his time allowed this.

Like all naturalists he learned to draw. But while his notes show beetles and birds, there are no planets on the pages. Although a superb observer, he had no concern with the physics of his equipment. His lively sense of invention was restricted to what a farm boy knew about — simple observatories, observing chairs (he worked in a furniture factory). No photometers for him.

He did like this life, he had done all the 19th-century things with tremendous success. And that brings up a point many will dispute. Farm boys were supposed to learn from books. Lincoln reading by candlelight was a powerful influence. And all the great naturalists he knew were also writers. Hence, he must be a writer, too.

I once asked him which he was first, a writer or an amateur astronomer. His reply was quick. He was fundamentally a writer. His telescope was just a tool to collect material just as when younger he had collected rocks from the glacial drift.

Until *Variable Views* started reprinting them, few of Peltier's friends knew that he had published several dozen articles in magazines. They were conventional nature study. "Mythical Horse Hair Snakes," which appeared in *Nature* magazine in 1933 is a sample.

His problem was that he had not written a hardcover book. In 19th-century rural America a book was an instant passport to lifelong social standing. That's how Thoreau and John Muir got famous. But astronomy books are hard to sell to publishers. He envied his friend Edwin Way Teale who had published many books, and laments, "(Teale) authoring a score or more books; I had made no trail at all."

But with the advent of his first book all that changed; he had arrived. The success spurred him to *The Place on Jennings Creek*, which is slowly carving a real niche for itself among the sensitive.

Peltier's work will last, for he learned writing from real craftsmen. Thoreau runs alongside him in his imagery and sentence rhythms. Any writer who describes the last ice age as, "That famous dabbler in real estate the Great Ice Age," was well trained. His prose is effortless and reads like a flowing brook, sparkling with unexpected glints.

Dr. Lewis Epstein once told the Astronomical Society of the Pacific that the chief job of the amateur astronomer has always been to insure the continuation of the attitudes that have made astronomy great. This is more important than making variable star or meteor observations, valuable as they are. My mail each month attests to the spur that Leslie's books, his example, and his personality have contributed to this influence.

Peltier that night did a good deal for my attitudes, and in return we invited him to continue with us to Maryland and the AAVSO spring meeting. He turned shy instantly and mumbled about his farm chores.

I turned to his sister Dorothy and asked, "We might mess up your kitchen maybe." Her reply was a smiling, "I got a new clothesline out back. You can use that." Peltier went quietly.

The small convention was ecstatic when Peltier thus came to his first convention. He liked it, too. It was his first chance to talk to observers he knew only by mail. All the way home to Delphos he kept smiling his small bright smile.

In fact, he was so pleased he then assaulted the heavens with redoubled efforts and in a few months — there was another comet discovered.

—*Walter Scott Houston*

CHAPTER 1
Trail's Beginning

A guidebook should start at the beginning of the trail. This is especially true when the pathway leads through unfamiliar territory. Right here, then, at the very outset of a starwalk which will wind all the way across the skies and through all the seasons of the year, I will assume that you have never passed this way before. You don't know the name or the location of a single star. For you, the comings and goings of the planets are a total mystery. The Moon's round face is but a spotty globe, and you have never seen a sunspot or watched the variations in the brightness of a star.

To you, the Sun, the stars, the planets may always have appeared to be quite changeless when contrasted with a warbler in a springtime treetop. Your view of the skies seemed just the same from day to day and night to night. That was an illusion. You were really watching an animated scene, though you saw it from so great a distance that no movement caught your eye.

In coming chapters I will direct your eyes and binoculars to a number of locations where you'll find abundant proof that the sky is a busy place indeed, both day and night. Through dark filters you'll see sunspots larger than Earth. Sometimes you'll watch them grow from day to day as they move across the Sun. At other times you'll see them fade and disappear or even circumnavigate that turning, gaseous globe. By night you'll watch a sunrise two weeks long as it slowly moves across the Moon, while, in season, you will time the traffic of four other moons about a distant neighbor planet. On other nights, you'll meet a class of restless stars that will fade or

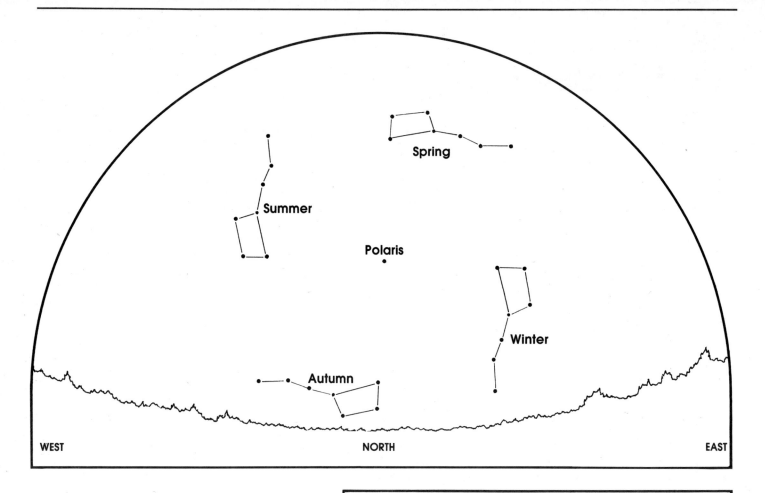

SEASONAL MOTION OF THE BIG DIPPER

Circling around the star Polaris, the Big Dipper moves across the sky from season to season. Observing at about 9 p.m., the constellation is east of true north on January 15; overhead on April 15; west of true north on July 15; and low on the northern horizon on October 15.

brighten by a thousandfold.

Yes, your "changeless" skies are filled with action, with mighty movements that dwarf anything we know here on Earth. And you can spot these changes in the skies and turn your binoculars on them with ever-growing interest and ease when you become familiar with their habitat, the stars and constellations.

We learn best by doing. The study of geology becomes more meaningful when we can hike to mountain, mine, or quarry and see the rocks in place. Botany takes on new life and interest when we plant a seed and watch it sprout and gradually develop roots and leaves and flowers. We learn to identify birds more quickly and effectively by stalking them at their activities in tree or shrub or meadow. In this same way, by actual involvement, the starry skies will cease to be a trackless maze. They will seem to come right down to meet you and become a part of your own neighborhood when you go out on your very first stargazing night and quickly find a certain star or star group.

And you can do exactly that!

The Big Dipper

On any clear night of the year — preferably when there is no bright moonlight — go out to some location that is fairly free of trees and lights and buildings and face the northern sky. Somewhere in this half of the sky dome, you are certain to see a group of seven conspicuous stars that are realistically arranged in the shape of a dipper. Appropriately, this formation is usually known as the Big Dipper, and the drawing on page 2 shows it just as you will see it in the northern sky at a convenient evening hour near the middle of each of the four seasons of the year.

If your first stargazing session is on a clear winter night — say around January 15 at about 9 p.m. — you will find the Dipper somewhat to the right of true north and apparently standing upright on its handle, as I have shown it here. Three months later — on April 15 at the same mid-evening hour — the Dipper can be found almost directly overhead. In July, it will appear in the west, and on October 15, you will find it low in the north — where the autumn mists of the horizon may make it difficult to see all seven of the stars.

By now you can clearly see that the Big Dipper seems to make a complete counterclockwise circle in the northern sky in one year's time. Oddly enough, it also makes this same circle in just one day! This is something you can easily check for yourself. If you look for the Dipper an hour or two after your 9 p.m. sighting on January 15, you will find that those seven stars are now noticeably higher in the sky. If you delay this second sighting a full six hours — to 3 a.m. on the morning of January 16 — you will find the Big Dipper high above your head in the April 15 position. Thus, the apparent daily movement of the stars from 9 p.m. to 3 a.m., or one-fourth of a day, equals the apparent seasonal movement from January 15 to April 15, or one-fourth of a year.

In watching the Big Dipper's changing positions you may already have noticed that it seems to be revolving slowly about a central point in the sky that is marked by a star of about the same brightness as the Dipper stars. This star is called Polaris, the North Star, for it is close beside the North Pole of the sky. Note, too, as you continue to watch this circling Dipper, that no matter at what hour of the night or at what season of the year you look, the two end stars in the Dipper's bowl point almost directly to Polaris. For this reason, these two stars are called the Pointers.

It is this circling of the Big Dipper about Polaris that demonstrates so clearly the apparent movement of the stars. I use the word *apparent* here, for the ceaseless circling that you see is not due to any actual movement of the stars themselves but is, instead, the effect produced by two separate motions of the Earth. It is the spinning of the Earth on its axis — from west to east once every twenty-four hours — that not only gives us night and day but also creates the illusion that the stars of the Big Dipper are revolving around Polaris once each day.

It is the Earth's other movement, the year-long journey it makes around the Sun, that causes the slow seasonal shifting of the positions of the stars. We on Earth are, in effect, going to meet the stars that lie before us in the east, and this forward eastward movement constantly changes our position in respect to all the stars, causing each star to arrive at its same position in the sky almost four minutes earlier each night. You can easily verify this if you watch some star that is rising in the eastern sky. If, when viewed from your fixed position, this star clears the rooftop of some building at exactly 10 p.m., you will find that it will emerge at 9:56 on the following night. These four-minute nightly gains add up to two hours every month and to twenty-four hours in a year, so that one year after your first sighting, that same star will come over that same rooftop once again at exactly 10 p.m.

In addition to this apparent east to west motion of the stars — which you can readily detect in half an hour of attentive watching — each star has a very real motion of its own. It may be at a speed of thousands of miles per minute, but the stars are so distant that the term "fixed stars" is often applied to them, for they do seem fixed in their positions in the sky in relation to each other. However, in the long view of time they are anything but fixed, and no stars show this proper motion, as it is called, more clearly than the stars of the Big Dipper.

The drawing at the right illustrates a chapter in the long life story of these seven stars. The upper sketch shows the Dipper as it appeared to early man some 100,000 years ago. Next, it's shown as we see it today. Below that is today's Dipper, with arrows showing the direction and amount of motion of each star over the next 100,000 years. Note that two of these stars, the upper Pointer star and the star in the end of the handle, are moving in a direction nearly opposite that of the other five stars. This suggests that these five all belong to a moving star stream and may have had a common origin. It would appear that here we are witnessing the meeting and passing of two opposing streams in a stellar flight that is certain to spell eventual doom for the figure of the Big Dipper. For in 100,000 years, as the bottom figure shows, the handle of the Dipper will be badly bent and its bowl will flatten out until the Pointers will no longer serve to lead the eye to bright Polaris.

Constellations

It was probably the early stargazers who dwelt in the valleys

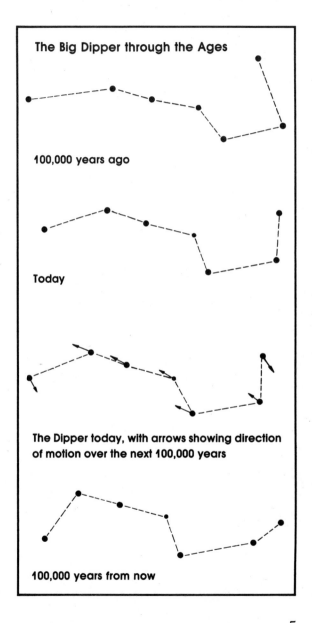

The Big Dipper through the Ages

100,000 years ago

Today

The Dipper today, with arrows showing direction of motion over the next 100,000 years

100,000 years from now

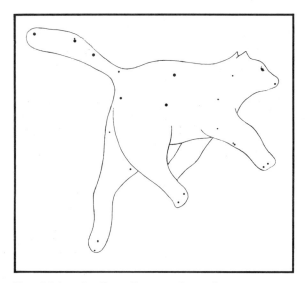

Ursa Major, the Great Bear, as the ancient stargazers saw it in the sky.

of the Tigris and Euphrates Rivers some 5,000 years ago who gave us our constellations or star pictures in the sky. Living in the outdoors as shepherds, farmers, desert nomads, and tribal warriors, they saw in the random groupings of the stars some resemblance to many of the familiar things about them. They peopled the skies with other herdsmen, hunters, and archers. They found in those early starfields their domesticated and wild animals such as dogs, bears, horses, and bulls. Later on, when the Greeks learned of these sky pictures and became involved in stargazing themselves, the folklore and mythology of their country was also hung in the celestial gallery in the form of a king, a queen, and many of their fabled heroes.

The constellations, fanciful though they were in outline and design, served a worthy purpose. They gave each star a family name as well as a home territory. I can't imagine how anyone could possibly learn the stars if each one was just an individual with no close relatives or other family ties. Even the most accomplished stargazer can't recognize stars that appear all by themselves. They must be in their proper setting, surrounded by their neighbor stars, to be identified correctly. By segregating the stars that were visible in the night skies of their home latitude into some forty-eight different groups or constellations, each depicting some familiar bit of folklore, those early watchers made the first step toward bringing scientific order to the stars.

The eighty-eight constellations of the entire sky are listed in Appendix V together with their abbreviations and a space for the date of your first sighting of them.

Ursa Major

The ancient weavers of the sky patterns placed the seven stars of the Big Dipper in the constellation known as Ursa Major, the Great Bear. As they saw it in the sky, the handle of the Dipper forms the bear's tail while the bowl extends well to the middle of the creature's back. One faint star marks the bear's nose while three pairs of stars outline the feet. Don't expect to see anything about this star group that in any way resembles a bear, for it simply isn't there. A modern attempt at rearranging all constellation outlines for greater realism shows the bear facing in the opposite direction with its nose marked by Alkaid, in the end of the Dipper's handle. But this bear is just as far-fetched as before, and the sky thereby loses some 5,000 years of history.

Fortunately, in our study of the skies it is not necessary to trace out any of these ancient patterns, for only in three or four cases is there any actual resemblance between the constellation figure and the earthly prototype for which it was named. It is of far greater importance that you become familiar with each star group as *you* see it. For example, you should be able to quickly locate the Big Dipper at any time of night and at any season of the year. Then you can use its stars to guide your eyes and your binoculars to any feature of interest in the entire constellation. Furthermore, as you have already seen, the Big Dipper also obligingly points the way to still other stars and constellations, such as Polaris in the handle of the Little Dipper.

CHAPTER 2
The Little Dipper and Polaris

To the nameless namers of the stars, the Little Dipper formed the constellation Ursa Minor, the Little Bear. Here as in the Great Bear, we find the figure of a dipper made up of seven stars and, again, four of these are in the bowl and three make up the handle — though in this case the handle curves up instead of down.

On ancient star charts, the tail of the Little Bear is virtually as long as the creature's body with its tip marked by a bright star, Polaris, which now stands near the north pole of the sky. Actually, the true celestial pole — the point exactly in line with the Earth's 8,000-mile-long axis — is two Full-Moon diameters distant from Polaris and lies on a line drawn from Polaris to the star Mizar in the handle of the Big Dipper. This means that Polaris, like all the other visible stars, circles around the true pole once every twenty-four hours. You can see this for yourself by taking a time-exposure photograph of the polar region some clear night. Your photo will show Polaris as a bright, circular arc whose center is the sky's true polar point.

Polaris may well become the first star that you enter in your life-list of celestial objects (see Appendix VI). It's a fine beginning, for it is easily the most useful star in all the night sky. For long ages, Polaris has served as man's guiding light in his nocturnal wanderings. The early Nordic sailors learned to steer their ships with this star as their navigating beacon; in later times, both pioneer and plainsman depended on the friendly reassurance of its gleam. Nineteenth-century poet William Cullen Bryant in his "Hymn to the North Star" pays this tribute to the star's importance:

> On thy unaltering blaze,
> The half-wrecked mariner, his compass lost,
> Fixes his steady gaze,
> And steers, undoubting, to the friendly coast;
> And they who stray in perilous wastes, by night,
> Are glad when thou dost shine to guide their footsteps right.

Today Polaris is still a star that everyone should know — be he sailor, scout, or stargazer — for whoever looks directly at it knows that he is facing north, that east is at his right, that west is on his left, and south is at his back. The beginning watcher will find that this star's steadfast glow can serve to guide him in his early exploration of the skies.

Polaris is also an aid in determining your own latitude on Earth's surface, for the north pole of the sky is the same number of degrees above the northern horizon as you are north of Earth's equator. For example, if you live in New York City, Philadelphia, Pittsburgh, Indianapolis, or Denver, you are roughly 40 degrees north of the equator. For you, Polaris would be 40 degrees above the north horizon line, or a little less than halfway up the sky from the horizon to the zenith, the point directly overhead. (The observational information in this book is oriented to stargazers living at this approximate latitude.) As you travel north, Polaris appears to climb higher and higher in the sky. At Minneapolis, which is at 45 degrees north latitude, the pole star is just halfway up the sky. At the North Pole, the celestial pole will be directly above you, in your zenith, with Polaris circling closely about it every twenty-four hours.

If you were standing at the North Pole, you could only move in one direction — south. You are already as far north as you can go and every point on Earth is south of you. Rudyard Kipling was skirting scientific truth when he wrote:

> For there is neither East or West,
> Border or breed or birth,
> When two strong men stand face to face,
> Though they come from the ends of the Earth.

His "two strong men" must have been standing on the poles, the ends of the Earth, for only at those points is there neither east or west.

The configuration of the Little Dipper is not nearly as conspicuous as that of the Big Dipper. In fact, in moonlit or hazy skies, your unassisted eye may have some difficulty in locating all seven of its stars. Only Polaris (α) and Kochab (β) are as bright as the stars of the Big Dipper.

The brightness of stars is of particular interest to stargazers. In the pages to follow, I will often call attention to a star's brightness, which is called its magnitude. This is not a measure of a star's size but of its relative brightness as we see it from Earth. For example, Sirius, the brightest of all the stars, appears so bright to us because it is the nearest of all the naked-eye stars visible from our latitude. On the other hand, Canopus, the next brightest star, though too far south for us to see, is some thirty times as distant as Sirius and owes its brilliance to its truly enormous size and hot surface. If we could view these two stars from some point in space midway between them, Canopus would greatly outshine Sirius.

There are about 6,500 naked-eye stars scattered over the entire sky. In the second century, the Greek astronomer Claudius Ptolemy divided them into six degrees of brightness: from the very brightest, which he called 1st magnitude, down to the faintest he could see, which became 6th magnitude. A light ratio of 2.5 was decided upon between each magnitude; that is, a 2nd-magnitude star is two-and-a-half times fainter than a 1st-magnitude star, and a 3rd-magnitude star two-and-a-half times fainter than a 2nd-magnitude, and so on.

Ptolemy's system of magnitudes is still used today. Its only complication lies in the 1st-magnitude division, which has such a range in brightness that this group really covers three magnitudes. You'll find the precise magnitude of all twenty-one 1st-magnitude stars listed in Appendix IV; note that it includes every star brighter than magnitude 1.5. Similarly, 2nd magnitude includes all stars between 1.6 and 2.5 magnitude. There are about sixty of these in the sky — three times the number of 1st-magnitude stars.

This same brightness ratio and the same three-fold increase hold true for the other magnitudes, so that we have about 4,500 sixth-magnitude stars over the entire sky — each one of which is one-hundredth as bright as a 1st-magnitude star.

The Little Dipper conveniently furnishes examples of each of these naked-eye magnitudes except 1st magnitude. Since these stars are all well up in the sky every clear night throughout the year, you can quickly refer to them to estimate the brightness of other stars. Each magnitude has its standard magnitude stars that are about midway in brightness between the magnitude limits. Thus Polaris, at magnitude 2.0, is a standard 2nd-magnitude star, while Kochab, at magnitude 2.1, is only slightly fainter.

Note that the four stars of the Little Dipper's bowl represent four different magnitudes. Three of them, Beta (β), Gamma (γ) and Eta (η), can be used as standard 2nd, 3rd, and 5th magnitudes, while Zeta (ζ) is a rather faint 4th magnitude. And the faint star just below the bowl is a close example of 6th magnitude.

Don't be discouraged if you have some difficulty tracing out all the stars of the Little Dipper. This isn't an easy figure to catch with the naked eye as it requires a sparkling, moonless night and a location free from interfering lights. But even under poor conditions, all these Dipper stars and many others even fainter are well within the grasp of your binoculars.

These first two chapters have shown you just how easy it can be to meet the stars and to learn to recognize them. You are now familiar with the naked-eye appearance of the two most important constellations in the sky: Ursa Major, with the seven bright stars of the Big Dipper whose Pointer Stars then led you on to Polaris and the Little Dipper in the constellation Ursa Minor. These two star groups can serve you well in your further exploration of the skies, for they are year-round beacons that shine throughout the night to orient and guide you to still other stars and constellations.

Your eyes will always be the most precious and the most versatile instruments at your command. True wide-angle lenses, your eyes take in a vast expanse that reveals, at a glance, the entire picture you are looking for. As automatic zoom lenses, they hold in constant focus both the near at hand and the far away. But, as you doubtless learned on your first field trip as a Scout or camper, your eyes could

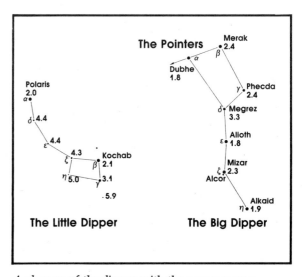

A close-up of the dippers with the proper names, Greek letter designations, and magnitudes of the brighter stars of both dippers.

often use more power.

This same need for added strength is experienced by those who follow star trails, for within these constellations and scattered all about the sky there are celestial objects of interest and charm that lie beyond the reach of the unaided eye. Hidden there are sparkling star fields, misty galaxies and nebulae, compact clusters of a hundred thousand suns, and restless stars that change in brightness from one night to the next. There, too, are double stars and colored stars, roving fellow planets, circling satellites, and the ever-changing face of our nearby neighbor Moon. Often there are streaking meteors and, more rarely, a passing comet or an exploding star. All of these are better seen through the probing powers of binoculars.

In coming chapters, I will discuss these attractions in detail and, with charts, show you where to find them in the sky. You'll find that the binoculars that have so enriched your daylight hours have some surprising talents that only show up after dark.

CHAPTER 3
Augmented Eyes

On any organized field trip you are likely to find a great variety of optical aids being used to bring the bird in the bush visually closer to the hand. These may be in the form of field glasses, prism binoculars, opera glasses, and sometimes even spotting scopes mounted on tripods.

Smallest, lightest, and least powerful of these accessories are the opera glasses that were once so popular with the theater-goer of a pre-TV era. I have seen some that were real gems of the optician's art, richly mounted in frames of mother-of-pearl. Seeing such treasures of long ago, I have often wondered what rare stage specimens and what *vox humana* warblers they may once have held in focus.

Opera glasses magnify about three times (3x), which means that objects seen with them appear to be three times closer than they actually are. At the opera, this would give you the advantage — though not the prestige — of a front-row-center seat, while on a bird walk it means that the songster in the treetop 30 feet away seems to be perching on the lowest branch only 10 feet away. For the stargazer, it makes all the Little Dipper stars stand out clearly, even on nights of haze or moonlight. They are helpful in hunting unfamiliar constellations, in following the brighter comets, or in tracing the dim outlines of the Milky Way.

My grandfather once had opera glasses that were truly the last word in compactness. They had been made in Germany and, except for the four lenses, were constructed entirely of celluloid. Two tubes that telescoped together formed the body of the instrument and

An Opera-Glass

Prism Binoculars

provided a means of focusing. The two large front lenses, set in celluloid frames that were hinged to the body tube, could be folded back against the body. The two eyepiece lenses were similarly mounted and folded forward. The whole affair collapsed to almost nothing and could easily be tucked into a vest pocket.

Such glasses are probably not being made today so, unless you already have a pair, I recommend that you invest in something better. Nor do I advise that you buy field glasses of the type sometimes seen on bird walks and nature trails. They are really only opera glasses of somewhat greater power. Glasses of this type are known as Galilean field glasses, for they use the same optical principle that was used by Galileo in his first telescopes.

That "something better" than either opera glasses or Galilean field glasses is prism binoculars. They come in so many styles, sizes, powers, field widths, and prices that their respective specifications should be carefully examined before making a choice.

Prism binoculars get their name from the fact that each tube or barrel contains a set of glass prisms that intercept the incoming light and pass it through a series of reflections that allow packing a longer focal length into the instrument without adding to the length of the barrels. This optical system was invented more than a hundred years ago by a French engineer named Porro, and the prisms that he devised still bear his name. In some modern binoculars another system is now being used in which roof prisms are employed. This results in a lighter, more compact instrument in which straight barrels are used rather than the offset barrels of the porro prism models. This type is also more expensive.

Somewhere on the body of the binoculars you will usually find a series of figures that tell the power of the glass and the diameter of the objective or front lenses. Most often these figures will read 7x35 or 7x50. In either case, the 7x means that the binocular magnifies an object seven times, while the 35 or 50 indicates an objective lens that is 35mm or 50mm in diameter. 7x glasses are a great favorite of birdwatchers — and with good reason. A power of seven is about the limit of magnification that you can hold steady in your hand. Glasses much stronger than this magnify every pulse beat, nervous tremble, and gust of wind. Plenty of light is admitted by 35mm lenses under average conditions, but near twilight or in deep woods the 50mm objective is superior — though it also adds to the weight, bulk, and expense of the instrument.

Powers of 10x and greater are justified for special uses such as for wildlife watchers whose quarry cannot be closely approached, i.e., deer in the forest, though here the use of a tripod or some form of steady mount will greatly help. If your stargazing studies require a larger image of the Moon or planets, then 10x is, of course, an advantage. Here, perhaps, you can devise a more permanent mounting in a suitable location and take the binoculars in after each use.

Among the finer features of prism binoculars is the broad, sharp field of view which they afford. This field dimension is often printed on the body of the instrument telling the width of the field (in feet) at a distance of 1,000 yards. Most 7x35 glasses have fields ranging in width from 367 to 420 feet. These are normal fields. However, there is another class known as wide-field binoculars that have fields more than 500 feet in width. No doubt they were originally designed for the sportsman, for they permit a more inclusive view of the playing field and the race track. In constellation hunting, this inclusive field of view is supreme since it allows a single field to hold the entire pattern of a star group or the flowing river of the Milky Way from shore to shore.

The diameter of the exit pupil of binoculars determines the amount of light received by your eye. The size of the exit pupil can be found by dividing the diameter of the objective by the power. In a 7x35 glass, the exit pupil is 5mm, which is ample in ordinary light as it is larger than the 4mm diameter of the pupil of your eye. 7x50 binoculars have a 7.1mm exit pupil, which is better in poor light and allows still more leeway in case you can't hold the binoculars perfectly steady. If your glasses have an exit pupil less than 4mm in diameter, you are not using your own eye pupil to its full capacity.

Coated optics are a decided improvement that older glasses lack. They consist of a thin coating of magnesium fluoride which reduces glare and reflections on internal surfaces. These coatings can be seen as reflected blue or amber colors when held at an angle to the light.

If you are buying binoculars, a great variety are available at prices ranging from $35 to $1,000. At the high end of this scale are the glasses offered by the same companies that make the more expensive cameras. There are, however, many mail order houses that carry good binoculars in the low-to-medium price range, say from $75 to $300, that will perform splendidly for you and bring just as much pleasure and as many happy observing hours as the high-priced luxury jobs.

I have owned two different pairs of binoculars. My first pair arrived some fifty years ago and, as I remember, cost about $15 — and these weren't the cheapest in the catalog! They were imports, as nearly all glasses are even today, but I made an unwise choice. They were 8x25mm, a combination that gave me a dark and narrow field with an exit pupil of only 3.1mm so that only one-fourth of my visual capacity was utilized. These glasses did have their good points, however. They were light in weight and small in size and, best of all, they fitted edgewise into a sturdy leather case that measured only 4½ by 6 inches. They could be brought into action using only one hand with about the same fast draw as a western outlaw with his six-gun.

At the time I was blissfully unaware of the faults of these binoculars. I saw a lot of birds with them, and with their help I trailed the nightly movements of some comets of that era. Today, those glasses have a place beside our kitchen window where they are always handy for a closer inspection of the visitors to our feeding station fifteen feet away.

When I bought my second pair of prism binoculars, I knew just what I wanted. I don't remember where I got those glasses, though I'm sure they were from one of the discount houses such as Joseph Hagn, John Plain, or Bennett's whose bulky catalogs of merchandise were to be found in every office. This time my selection was 7x35 wide-angle binoculars, and after thirty years of use, I still have found no fault with them. The exit pupil is, of course, 5mm, while their wide field — 531 feet at 1,000 yards — shows a sky just over 10 degrees in diameter so that the entire bowl of the Big Dipper can be seen in one field. They have coated lenses fore and aft and show sharp star images to the very edge of the field. They weigh just 24 ounces — lightweight compared to the 7x50 size.

These glasses, made in the more streamlined American style, were imported from Japan, though I can find no maker's name on them other than a trade name — SEEFAR. They cost $35 at the time of purchase. All other things being equal, my own preference favors 7x35mm over 7x50mm. Not only are the former less expensive, they are lighter and less bulky to handle and seem to have somewhat better definition than some of the 7x50s I have tried. For my own stargazing, in skies that are becoming brighter every year, I find that the darker fields of the 35mm give sharper contrast than the bright sky backgrounds of the 50mm.

Before purchasing any binoculars, new or used, make a thorough quality test. Most important, check for image quality. Stars, for example, should be clear and sharply defined over the entire field of view. Both barrels of the instrument should be perfectly aligned so that they form a single bright image with no overlapping. Be sure that the barrels can be adjusted to the correct interpupillary separation of your eyes.

You can test the quality of the glass prisms in your binoculars by holding the instrument a few inches in front of your eyes and pointing it to the open sky. With the best quality glasses, the exit pupils will show as two circular disks of light. With cheaper instruments you will see a square or oblong within the bright circle of each pupil.

If the power of the binoculars has not been stamped on the body or has been defaced, you can determine it yourself. Aim your glasses at a brick wall or at the side of a frame house, then look through one barrel with one eye while your other eye surveys this same target outside the barrel. How many bricks or siding boards, as seen with the naked eye, does it take to fill the space of just one brick or board as seen through the glass? If seven are required, you have 7x binoculars.

The Big Dipper will help you estimate the angular width of your instrument's field of view using 10 degrees as the distance across the top of the bowl, 7 degrees across the bottom, and 5 degrees between the Pointers.

Now, with your natural curiosity, patience, and perseverance, you are ready to make the acquaintance of the night skies of every season of the year. The more familiar you become with the stars and star groups, especially with their naked-eye appearance, the more readily you can train your searching glasses on the many fascinating objects hidden there.

As an introduction to these brighter stars and constellations, I have drawn a series of maps and charts that graphically picture the skies as seen from our mid-northern latitudes. They should serve to direct your eyes and then your binoculars to the inviting paths that await your exploration. You have already met three trusty guides to lead the way: the Big Dipper, the Little Dipper, and Polaris.

CHAPTER 4
Maps of the Sky

In an earlier chapter you became familiar with the figure of the Big Dipper. You saw, too, just how easy it was to locate the North Star, Polaris, in the Little Dipper merely by following with your eye the line projected by the two bright Pointer Stars.

By this same simple method you can become acquainted with all the other stars and constellations. Your skies have many other pointer stars and reference lines that will introduce you to a neighbor star or star group and it, in turn, will point to still another until, before long, you will be familiar with all the brighter stars and constellations that cross your skies throughout the year.

In the following series of sky maps and key charts you will find the stars and constellations pictured as they appear at the season of the year when they are in the best position for viewing. Because there is often haze and sky-glow as well as trees and buildings near the horizon, the best observing time for most of us comes when the object of our search is on or near the meridian, an imaginary line that passes through the north and south celestial poles and our zenith. A star on the meridian is at its highest point in its daily travel from east to west.

Each sky map and key chart shows the sky as it appears on the semi-monthly dates at the Standard Time hours listed. For daylight-saving time add one hour, and for dates and hours other than those given, the appearance of the sky will be a close approximation to the nearest given date and evening hour. The sky maps, on the right-hand pages, show the nighttime appearance of the skies with some landscape features added for scale and realism. On the left-hand pages directly opposite the sky maps are the key charts. Along each margin are printed the compass directions covered by each chart. Here the stars are plotted exactly as they appear on the sky maps, but now the stars, constellations, and various features of interest have been labeled with their names. The Latin names of the constellations appear in larger type, while the proper names of many of the brighter stars and the Greek letter designations are given in smaller type. Also in the key charts many of the more prominent stars have been connected with solid lines in order to make a definite design. You will note how much more quickly your eye picks up the figure of the Big Dipper, for example, when its seven stars are connected in this way.

Throughout the book I have given the English names of the Greek letters. The entire Greek alphabet is as follows:

Alpha	α	Iota	ι	Rho	ϱ
Beta	β	Kappa	κ	Sigma	σ
Gamma	γ	Lambda	λ	Tau	τ
Delta	δ	Mu	μ	Upsilon	υ
Epsilon	ε	Nu	ν	Phi	φ
Zeta	ζ	Xi	ξ	Chi	χ
Eta	η	Omicron	ο	Psi	ψ
Theta	θ	Pi	π	Omega	ω

There are two sets of maps and charts for each season of the year. The first set for each season shows the half of the sky you see when you are facing north; the second set depicts the remaining half of sky as it appears when you turn around and face due south. Inasmuch as the sky presents the appearance of an inverted bowl, it is quite impossible to show it on a flat page without some distortion around the edges. This shouldn't cause any problems, as virtually all of our serious observing will be confined to the mid-regions where there is the least atmospheric interference and where distortion is at a minimum. The letter Z, high up on each key chart, marks the location of your zenith, the point in the sky directly above you, a point in the sky that is yours alone.

CHAPTER 5
The Winter Stars

Facing North

In the sky map on page 23, which covers the northern half of the sky, are two star figures that you have already identified. First you located the Big Dipper with its two Pointer Stars; this then introduced you to the second figure, the Little Dipper and the star Polaris.

To the south and east of the Big Dipper's bowl you will find a number of fainter stars and these, together with the seven stars of the Dipper, make up the constellation Ursa Major, the Greater Bear. The best known star of this large constellation is Zeta, or Mizar, the middle star in the Dipper's handle. Even a casual glance at this star will show that it is not alone, for a 4th-magnitude star named Alcor travels through space right beside it, making a naked-eye double star. There are thousands of double stars scattered about the sky, but Mizar is more than doubly famous. Not only was it, with Alcor, probably the first known naked-eye double, but Mizar itself was the first double star discovered with a telescope. This happened in 1650 — some forty years after the invention of the telescope — which would seem to show that the early sky watchers were more concerned with the Moon and planets than they were with stars.

But there is even more to Mizar than meets the eye in any telescope. The brighter of the telescopic pair later proved to be a spectroscopic double star and, again, it was the first one ever found. Too close to be detected in a telescope, the periodic doubling of its spectral lines showed Mizar to have a faint companion star revolving about it in 20 days at a distance of about 36 million miles — less

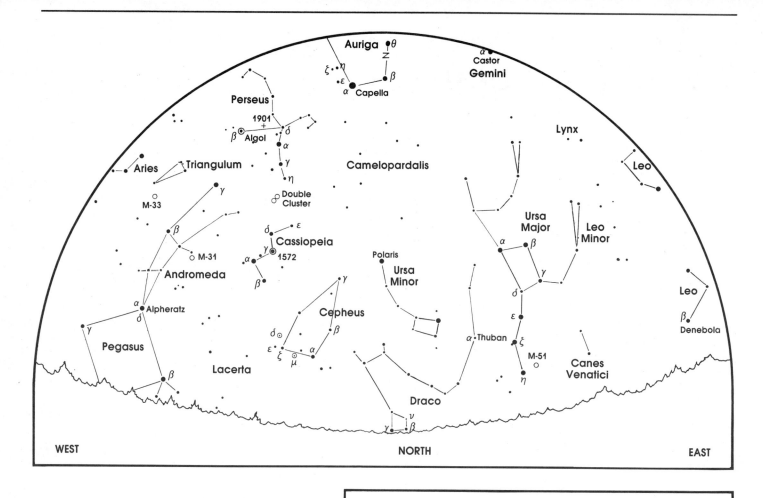

CONSTELLATIONS OF THE NORTHERN WINTER SKY

ANDROMEDA   CAMELOPARDALIS   CASSIOPEIA
CEPHEUS   LEO MINOR   URSA MAJOR   URSA MINOR

THE NORTHERN WINTER SKY AS SEEN ON:

DECEMBER 15 AT 12 A.M.   FEBRUARY 1 AT 9 P.M.
JANUARY 1 AT 11 P.M.   FEBRUARY 15 AT 8 P.M.
JANUARY 15 AT 10 P.M.   MARCH 1 AT 7 P.M.

THE WINTER STARS

than half the distance between Earth and the Sun.

There are two general types of double stars: optical doubles and binary systems. Optical doubles are those in which two stars happen to be nearly in the same line-of-sight. This makes them appear to be side by side and almost touching when, in reality, one of the pair is enormously farther away from us than the other. Since there is no physical connection between the two stars, it is perhaps stretching a point to even call them double stars.

Binary stars, on the other hand, are true doubles, for they form a system of two stars revolving about their common center of gravity. As in the case of Mizar and its faint companion, binary stars are usually relatively close to each other. Contrary to what you might expect, about one-third of all the stars, the whole sky over, belong to binary or multiple star systems — all of them visible proof that gravitational attraction is indeed a universal law.

In our exploration of the skies, we will occasionally view double stars of both types, but I will comment only briefly on the more exceptional of these for, in general, they aren't rewarding objects in binoculars and there is little that we can learn from them.

Well to the west at this season of the year, the Milky Way sweeps upward from the horizon to the zenith. The lower part, however, is often contaminated by the haze and heavy atmosphere near the horizon and by the interference of trees and buildings, unless you can get out into open country away from such distractions. I realize that few are free to do this except on rare occasions, and it is for this reason that I will discuss the stars and constellations at the times and seasons when they are well up in the sky where they can be seen through our thinnest atmosphere.

In the western sector of this northern winter sky some stars of 2nd magnitude can be seen lighting up the Milky Way, but they have already passed their prime position in the sky, and so they will be described later on when they are rising in the east as autumn stars.

There is a dearth of 1st-magnitude stars in the northern sky at this time of year. Only one such star, bright Capella, can be found, and it is now so nearly overhead that it can be more comfortably observed when it will be climbing in the eastern sky in autumn.

Perhaps this scarcity of northern stars is just as well right now, for as you turn and face the south the year's most brilliant skies are on display.

Facing South

If you have ever wondered what to do on long winter evenings, the southern skies provide the answer. Here, scattered about from east to west and from zenith to horizon, are no less than eight 1st-magnitude stars — more than half the total number of these stars that you can see from mid-northern latitudes. Nowhere in the heavens will you find another such glittering array, for six of these bright stars are now grouped near the center of the winter stage.

Orion

The most brilliant of all constellations is Orion, the Hunter, whose bold figure now stands upright in the southern skies just west of the meridian. With the exception of the Big Dipper, Orion is the best known star group in the sky and rightly so, for it includes two stars of 1st magnitude, four of 2nd magnitude, as well as a host of fainter stars. In my early atlas of the skies, Orion is shown as a giant who holds an uplifted club in his right hand, while from his left arm hangs a shield made from a lion's skin. Before him stands Taurus, the Bull, with lowered head and gleaming reddish eye. Actually, it doesn't take much imagination to see this age-old combat in the skies, for the figures are well outlined with stars.

Marking the right shoulder of the giant, Betelgeuse is the only 1st-magnitude star that varies in its brightness. This irregular fluctuation amounts to only half a magnitude and probably is a result of the star's extreme age. Betelgeuse is thought to be a dying sun, one of the class of stars known as red giants, an especially fitting title in this case since it is a giant in actual size and its red hue is quite apparent to the naked eye. In 1920 Betelgeuse became the first star whose diameter was directly measured. This was accomplished with the aid of an instrument called an interferometer attached to the 100-inch Mt. Wilson telescope, then the largest in the world. This measurement gave Betelgeuse a diameter of 260 million miles — truly a giant when compared to our Sun, whose diameter is 860,000 miles.

CONSTELLATIONS OF THE SOUTHERN WINTER SKY
CANIS MAJOR   CANIS MINOR   COLUMBA   ERIDANUS
GEMINI   LEPUS   MONOCEROS   ORION
PUPPIS   TAURUS

THE WINTER STARS

THE SOUTHERN WINTER SKY AS SEEN ON:

DECEMBER 15 AT 12 A.M. FEBRUARY 1 AT 9 P.M.
JANUARY 1 AT 11 P.M. FEBRUARY 15 AT 8 P.M.
JANUARY 15 AT 10 P.M. MARCH 1 AT 7 P.M.

THE WINTER STARS

Diagonally opposite Betelgeuse in the figure of Orion is Rigel, the constellation's brightest star. It, too, can only be talked of in superlatives, for it is one of the most remote of the visible stars as well as one of the very hottest of them all. Its distance is given as 1,300 light-years, meaning that light, which can encircle the Earth seven times in a single second, requires 1,300 years to reach us from this star! Yet even at this incredible distance, Rigel ranks seventh in order of brightness of all our stars. It would take the combined light of 150,000 Suns to equal that of Rigel.

Midway between Rigel and Betelgeuse, three 2nd-magnitude stars form a nearly perfect line just three degrees in length. This is Orion's Belt, which is referred to in the Book of Job. So striking and unusual is this three-starred figure that it serves as a means of identification for the constellation. Mintaka, the upper star of the Belt, is exactly on the equator of the sky. Therefore, if you were standing at our North Pole, this star would be precisely on your horizon, where it would mark out the southern boundary of your circling hemisphere of stars. Mintaka is a double star which 10x binoculars will split, but it still conceals its 7th-magnitude companion from my 7x glasses.

Hanging downward from the Belt is Orion's Sword. Even with your naked eye, you will note a haziness about the central star of this small shape. As the poet Tennyson writes of its appearance in "Merlin and Vivien":

> A single misty star,
> Which is the second in a line of stars
> That seem a sword beneath a belt of three.
> I never gazed upon it but I dreamt
> Of some vast charm concluded in that star
> To make fame nothing.

That "single misty star" is the Great Nebula of Orion, the finest of its type in the entire heavens. Here is a vast, formless cloud of incandescent gas and dust which is illuminated from within by a number of hot, blue stars embedded there. Binoculars greatly increase the visible size of this cloud in which only its brightest star, Theta (θ), can be seen with low power.

Designated as M-42, the Great Nebula is some 1,000 light-years from us, and it has been reckoned that light would require 15 years to cross this nebulosity from side to side. Oddly enough, Galileo, who made so many first sightings with his new telescope, missed the Great Nebula completely, though it was well within the powers of his tiny optic tube. More than two hundred years after Galileo's time, M-42 was watched by stargazer Lord Tennyson, who often "gazed upon it" with his own two-inch telescope.

Orion is a constellation of so many visual delights that you won't be limited by the dates and hours that I have listed here. They are for the times when he makes his *best* appearance, high up in a dark sky and well above the haze of the horizon. As you get to know this sparkling giant, you will come to watch for him as he rises in the east at about 10 o'clock on a frosty night in late October. As the weeks go by, you can follow his slow march across snowy winter landscapes to his final setting in the balmy twilights of late April.

Taurus

Sighting upward along a line drawn through the three Belt stars of Orion leads the eye close to rosy-red Aldebaran in the constellation Taurus, the Bull. This 1st-magnitude star represents the gleaming eye in the lowered head of the legendary figure. Here, within a small sky area, you'll find samples of three bright star colors. Through your binoculars, look long at rose-colored Aldebaran, then swing to red Betelgeuse, then to blue-white Rigel, then, finally, back again to Aldebaran. It will seem like you are viewing a display of garnets, rubies, and sparkling diamonds.

Aldebaran glows amid a star figure shaped like a letter V that forms a scattered open cluster called the Hyades. It is known to be a moving cluster, for all of its principal stars, with one exception, are speeding through space as a group. This suggests that, at some remote time and place, they may have started out together. The cluster is 140 light-years distant from us, and is moving in the direction of Betelgeuse in Orion.

Because of his superior brightness, you might assume that Aldebaran is the leader in the Hyades hegira, but he is not even a member of the clan. Actually, Aldebaran is somewhat nearer to us than to the cluster, which merely furnishes a sparkling background for the Bull's-eye star. Your naked eye can readily make out a couple of wide doubles in the outline of the V, and binoculars add scores of brilliant objects to the glittering field.

Curving upward from the lowered head of Taurus, his horns are tipped with two bright jewels, 3rd-magnitude Zeta and 2nd-magnitude Beta (β) or El Nath, which is also used to complete Auriga's pentagon. Near Zeta is M-1, the Crab Nebula, the hazy remains of a star that exploded in A.D. 1054. It was number one in a list of 103 diffuse objects catalogued by Charles Messier, the great comet hunter, in 1781. You may just glimpse it with a low-power telescope as a fuzzy starlike speck. It is properly a telescopic nebula, but there are many other Messier objects (such as Orion's M-42) that are within the range of your binoculars, and we will locate and discuss them as they come into view.

I have saved the best of Taurus until last, and now it seems almost a waste of time and effort to tell you where to locate it in the sky, for I have yet to find the person who has never seen the Pleiades or who didn't express amazement and delight at their appearance in binoculars. It is the finest open star cluster in the entire heavens.

Except for the Sun and Moon, this group is the first sky object that I learned to recognize. As a child of five, I first met the Pleiades. I still turn my glass on them whenever I am in their neighborhood and I still feel their charm and fascination. This is never more pronounced than when they make their first appearance of the season, like a little misty cloud just above the treetops on an early evening in October.

This small group is often called the Seven Sisters, and six of them are easily seen with the naked eye, even with some moonlight present. On the clearest of nights, I have often seen nine stars here, though some earlier observers have reported as many as fourteen naked-eye stars in the cluster. Binoculars, of course, reveal dozens more, though I can find no colored stars among them. This isn't remarkable, as the Pleiades — like the Hyades — form a moving cluster, so a common star type should be expected.

The Pleiades lie only some five degrees north of the ecliptic, and therefore the Moon, in its travels around the Earth, will occasionally move between us and the cluster causing some of its stars to be occulted, or hidden from us by the body of the Moon. When conditions are favorable, several of these occultations can occur in the course of a few months. I last saw this happen in late 1971 and 1972 when six of these blackouts took place. A repeat series is scheduled for 1987. If you are fortunate enough to witness one of these performances, you may be surprised to find that the Moon, which can seem so large when near the horizon, fits quite easily into the bowl-shaped figure of the little cluster. This entire group of stars is festooned with wisps of the nebulous matter from which its stars were formed. Its wraithlike aspect shows up best in long-exposure photographs, though you can sometimes get vague glimpses of it with low-power binoculars.

Eridanus

Close beside bright Rigel in Orion lies the source of Eridanus, the River, a name doubtless suggested by the meander of a stream of rather faint naked-eye stars that seems to first flow westward, then turns and drops to the horizon to finally end in the gleam of 1st-magnitude Achernar, a star too far south for us to see in mid-northern latitudes. There is little about this river to encourage any fishing with binoculars.

Lepus

Just beneath Orion crouches Lepus, the Hare. The ears of this timid creature appear as two wide pairs of stars below Rigel. Gamma (γ) is an easy double, magnitudes 3.8 and 6.4, that is 95" (seconds of arc) apart. The globular cluster M-79 appears as a bit of haze in your binoculars. To the west, in the location marked R, is the famous variable star R Leporis. This deep crimson star will be described and charted in a later chapter on variables.

Columba

Midway between Lepus and the south horizon is Columba, Noah's Dove. There is nothing of interest for our binoculars in this small four-starred figure, but I will mention it here to give you one more species to add to your Christmas bird count!

Canis Major

If a straight line from Aldebaran to Orion's Belt is followed onward for a similar distance, it will point out Sirius, the Dog Star, in the constellation Canis Major, the Greater Dog. Sirius is by far the brightest star in the sky with a magnitude of –1.5, which makes it twelve times brighter than Aldebaran. It is also nearer to Earth than any of the other naked-eye stars visible from our latitude, and this is largely responsible for its great brightness. Light requires only 8½ years to come to us from Sirius, and this — compared to the 1,300-light-year distance of Rigel — puts the Dog Star almost on our doorstep.

Sirius is a binary star with a most remarkable companion star that circles about it in a period of 50 years, but without a telescope of fair size, it is usually drowned in the bright glow of the primary body. Scarcely larger than three times the diameter of Earth, this tiny star is composed of material so dense that a teaspoonful of it would weigh a ton. If you weigh 150 pounds here on Earth, you would weigh 2,000 *tons* on the Pup, as this little companion of the Dog Star is often called. This miniature star is classified as a white dwarf. It probably is the remaining central portion of a once-large star which has used up all its hydrogen and then collapsed around its heavy nucleus.

To me, Sirius seems to flicker more than any other star. It is a white star, but because it is always rather low in our sky, thick layers of atmosphere cause it to throw off flashes of vivid color that are beautiful to watch in binoculars.

Canis Major contains four 2nd-magnitude stars which by themselves would make this a sparkling constellation, were they not so overwhelmed by brilliant Sirius. The ancients believed that when the Greater Dog rose with the Sun in late summer, the heat of its stars, added to that of the Sun, brought on the sultry period known as "dog days."

Look at the open cluster M-41 just below Sirius. It can be seen with the naked eye on clear, moonless nights; your binoculars will expand it into a lovely company of glittering stars.

Puppis

You may notice a few 2nd- and 3rd-magnitude stars lying between Canis Major and the southern horizon. They belong to the constellation Puppis, the stern of the ship Argo of Jason's legendary Argonautic Expedition. Most of the remainder of the ship lies below the horizon and can't ever be seen from our latitudes — which is unfortunate, as this excludes our seeing Canopus, the second brightest of all the stars.

The Milky Way flows through Puppis, and this makes fine sweeping for binoculars, with many rich fields and open clusters in the northern part of the region. Of these, M-46 and M-93 are the most rewarding, the latter having a diameter of about half a degree and containing numerous resolvable stars. The star Pi (π) lies in a fine field with bright companions.

Canis Minor

The first bright star east of Betelgeuse in Orion is Procyon, in the constellation Canis Minor, the Little Dog. Appropriately, it is a small constellation. Just above Procyon is the 3rd-magnitude star Beta with its two small attendant stars making a tiny triangle. Like Sirius, Procyon is a neighbor of our Sun, only 11 light-years distant. Oddly enough, it also has a faint white dwarf companion star circling about it in a period of 40 years.

The star 14 Canis Minoris is an optical triple star, which you can catch in binoculars when conditions are just right. Its magnitudes are 5.5, 8 and 9. See the key chart for the location of 14 CMi.

Monoceros

Within the triangle marked by Betelgeuse, Sirius, and Procyon lies most of the constellation Monoceros, the Unicorn. Even though the winter Milky Way runs directly through here, there are no bright stars to attract the eye, and the outline is vague and difficult to trace. There is, however, some rich sweeping here for binoculars, and some interesting clusters may be found. Of the latter, look at NGC-2244 clustered around a 6th-magnitude star and the open cluster M-50 about midway between Alpha and Beta.

Gemini

Standing to the north of Procyon, you will find the two bright stars, Castor and Pollux, in the constellation Gemini, the Twins. There is much of interest here. In mythology, Castor and Pollux were twin brothers who accompanied Jason on the Argonautic Expedition. Although these two stars appear in the sky to be somewhat closer together than the Pointers in the Dipper, there is no real relationship between them, for Castor is some 14 light-years more distant from us than Pollux. Castor, in fact, has no need for any family ties with Pollux; he is the brightest member of a family of his own — a system of six stars revolving about their common center of gravity. Many years ago, I could readily divide Castor and his 3rd-magnitude companion with my two-inch spyglass, but they have gradually drawn closer and this is no longer possible. There is a noticeable difference in brightness between these twins: Pollux is a faint 1st-magnitude star, while Castor is a bright 2nd magnitude.

Eta (η) Geminorum is classified as a red giant, an old star similar to Betelgeuse in Orion. It also varies from 3rd to 4th magnitude in a period of about eight months. The star Zeta is a fine double in binoculars and is a variable as well, with a range of half a magnitude in about 10 days. At its brightest it compares with Delta (δ) and Lambda (λ).

The star cluster M-35 is bright enough to be seen with the naked eye on clear, moonless nights, though binoculars will add greatly to its interest and beauty. Just west of M-35, a 5th-magnitude star marks the spot occupied by the Sun when it is farthest north at the beginning of summer (June 21). It was very close to this same spot in Gemini that, on the night of March 13, 1781, the English musician and amateur astronomer William Herschel, using a homemade reflecting telescope, came upon an object that showed a tiny disk rather than a starlike point. At first he thought it was a comet, but later observations showed the broad elliptical orbit typical of a planet. It eventually was named Uranus — the first new member to be added to the family of the Sun.

Gemini was once again to figure in astronomical history, for early in 1930, the tiny planet Pluto was found — this time with photography — near the 4th-magnitude star Delta Geminorum.

This ends our survey of the winter stars. Those constellations that you may have noted far to the west on the winter sky maps will all be treated later as autumn stars, after they have risen in the east and are well positioned for observing higher in the sky. The stars along the eastern border of the southward-looking winter map will become the stars of spring and, together with the northern skies of the vernal season, they now claim our full attention.

CHAPTER 6
The Spring Stars

Facing North

The sky maps and key charts on pages 38 and 39 show the skies as they appear three months later than the maps and charts of the winter season shown in Chapter 5. Inasmuch as one-fourth of a year has passed since then, the stars pictured here have made one-fourth of a revolution about the polar point. The Big Dipper, which in winter stood on its handle directly east of Polaris, is now almost overhead and at its highest point in the sky. Of course, all the other stars have also pivoted a quarter turn around the pole.

Now, low in the eastern sky, stars that will light our summer nights are rising, while other stars that we viewed high in our winter skies are now setting in the west. The Milky Way that was so bright before now lies all along the northern horizon and often is completely blotted out by haze. Only one star of all those pictured seems to have held its place. That star, of course, is Polaris.

But even *it* has not always stood so close beside the pole as it does today — nor will it always remain so in the future. All things change with time, and pole stars are no exception. If you could have stood somewhere upon a hilltop and watched, as in a time-lapse sequence, the age-long pageant of the slowly changing skies, you could have witnessed, as the millennia marched by, an awesome spectacle that no earthly eye has ever seen — the migration of the stars.

Just as most of our familiar birds fly north or south according to the seasons of the year, so do the stars, as seen from Earth, move north and south when prompted by the passing of another and much longer year — a year 25,800 Earth-years long that is sometimes referred to as the Great Year.

Both of these mass movements, of birds and of stars, are in obedience to the dictates of the polar axis of the Earth. It is the 23½-degree tilt of this axis that gives the Earth its seasons and thus provides the birds a motive for migration. It is another feature of this axis, a slow wobbling motion or gyration, that causes the migrations of the stars.

The Earth is not a perfect sphere. It is like a ball that is slightly flattened at its poles and bulged out at its equator. This distortion is sufficient to make its equatorial diameter 27 miles greater than its polar diameter. This equatorial bulge is tilted 23½ degrees to the ecliptic plane in which the Earth revolves about the Sun and, in obedience to the law of gravity, both the Sun and Moon are unceasing in their efforts to pull this encircling bulge into their common plane. Fortunately for birdwatchers and stargazers and all who appreciate the changes that the varied seasons bring, the rapidly spinning Earth acts as a gyroscope and is able to resist this pull — except for such side effects as the wobble of its axis.

At present, our polar axis points northward toward the star Polaris. It has been our pole star for hundreds of years, though its reign in that capacity is only temporary. The slow gyration of the axis causes it to point to a long succession of different pole stars that lie near a circle 47 degrees in diameter that the wandering pole traces out upon the sky. Any fairly bright star that lies near this circle serves as the pole star for a portion of the Great Year. The drawing on page 40 shows this circling path among the past and future pole stars of the northern sky.

I have often thought how fortunate our present-day Boy Scouts and Girl Scouts are to have such a conspicuous pole star as Polaris, one that they can find so easily by just sighting along the Big Dipper's Pointer Stars. Five thousand years ago when the Pyramids were being built, the pole star was Thuban in the constellation Draco. This star is much fainter than Polaris and is located about midway between Mizar in the Big Dipper and Kochab in the Little Dipper's bowl.

Thuban does serve well to illustrate the north and south

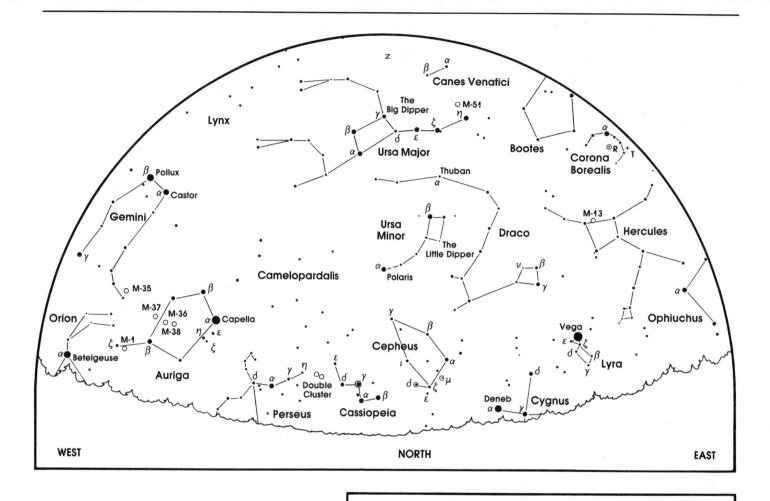

CONSTELLATIONS OF THE NORTHERN SPRING SKY
CAMELOPARDALIS CEPHEUS DRACO GEMINI
HERCULES URSA MAJOR

THE SPRING STARS

THE NORTHERN SPRING SKY AS SEEN ON:

MARCH 15 AT 12 A.M.   MAY 1 AT 9 P.M.
APRIL 1 AT 11 P.M.   MAY 15 AT 8 P.M.
APRIL 15 AT 10 P.M.

THE SPRING STARS

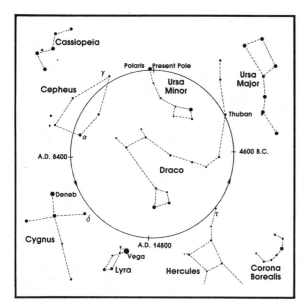

Pole Stars of the Great Year

migrations of the stars. In that long season of the Great Year when Thuban reigned as the pole star of the sky, the Southern Cross was on its northern migration flight, rising high enough above the southern horizon to be seen from as far north as Quebec, in Canada. Today it has flown south again and, in this country, the entire Cross can be seen only from the southern tips of Florida and Texas.

The true pole now is a little more than two Moon-diameters distant from Polaris and is slowly drawing closer to it. In the year 2095, it will be at its nearest — less than one Moon diameter away. The pole will then slowly move away in a counterclockwise direction. For long, long ages it will creep among the stars of Cepheus, first passing 3rd-magnitude Gamma Cephei around the year 4000 at about half the distance between the Pointer Stars. Then, a couple of millennia later, Alpha Cephei, or Alderamin, will become another distant pole star.

Some 12,000 years from now, the polar point will have completed almost half its circle in the sky, and blue-white Vega will be the brightest of all pole stars, even though it will be some five degrees distant from the true pole. The skies themselves will not have changed much in that time, save in their relation to the pole. We still would recognize all our familiar stars and constellations. The Big Dipper still will be a dipper, and Castor and Pollux, the Twins of Gemini, will still move side by side across the sky.

However, with the polar point near Vega, the habits and the seasonal associations of the stars and constellations will have vastly altered. No longer will Polaris and the Big Dipper be circumpolar, but instead they will rise and set. No longer will the Twin Stars ride high above our heads on winter nights. They will then be features of late summer evenings and will cross the sky just above the southern horizon. Missing completely from those distant future skies of our mid-latitudes will be Sirius, the brightest star of all. The dazzling Dog Star, along with blue-white Rigel and the Belt Stars of Orion, will then be on their southern migration flight.

Just as we look forward to the arrival of the springtime birds about our homes, so, too, is it comforting to know that in due time these migrating stars will all return to what, for us, seems like their proper seasons and home territories. The periodic migration of the stars is a vast cycle far more certain and predictable than the coming of the swallows to Capistrano or the mid-March meeting of turkey buzzards at Hinckley Ridge, Ohio.

Though no human eye can note its flow, the passing of the Great Year with its migrating stars is a repeating process. At some distant future date, the polar point will creep past Thuban then, ages later, the star Polaris will slow its spiral circling to once again become the Pole Star.

Facing South

Springtime, beyond a doubt, is the most exciting season of the bird-watcher's year. It is these three months — March, April, and May — that bring the birds back from the south again. Through many of these early days, the winter-weary eye is searching for the first sight of a robin hopping on the lawn, while all the time the ear is bent to catch the first trial birdsong tune-up notes that, as the days go by, will be augmented to a full-fledged chorus.

Spring also is a time of parting. With the first mild days, the friendly juncos that have been our guests all winter long begin to gather together in small twittering groups, as though discussing something of tremendous import. Soon, with a flash of white tail feathers, they fly to nearby evergreens where the small talk is resumed. The next day they are gone. Their vernal visa has expired, and they are off to their summer breeding grounds far to the north. But we feel no sadness of farewell when they depart, for we know, with certainty, that long before the first November snowfall they'll be back to spend another winter season with us.

The stars are even more responsive than the birds to the never-ending succession of the seasons. The rising of the stars in the east and their setting in the west is so punctual that the world's chronometers are set according to their crossing of the sky. Already you can see this relentless tide in action.

Go out just after dark on a mid-April night and look well down in the western sky. There you will see, sinking lower night by night, the entire coterie of bright stars that you watched on midwinter nights when they were high up in the southern sky. As you view this farewell party on succeeding nights, you will note that Rigel in Orion is the first to go, to be followed quickly by the Pleiades and then Aldebaran in the Hyades. Next to leave is bright Sirius who, as he nears the horizon, often shows a final flash of green or red. Then Betelgeuse, Procyon, Capella, and Pollux disappear in order as they drop behind the Earth.

As with the juncos, it is *au revoir* but not goodby to these bright winter stars that have left the western sky, for nothing is more certain than their rising in the east in the frosty nights of the coming autumn season. Again, as with the winter birds who left us when other travelers were arriving almost daily from the south, so too the departing winter stars have now been replaced by a lively company of spring stars and constellations. They now invite our close inspection, for often they will tax the powers of our probing binoculars.

Leo

Just to the west of the meridian and well up in the sky on a mid-season night, you can't miss 1st-magnitude Regulus in the constellation Leo, the Lion. According to the older star charts, Regulus marked the heart of the Lion, but for us it gleams at the end of the handle of the Sickle. As with the Big Dipper, this is a realistic figure.

With your binoculars note the difference in color between white Regulus and deep yellow Gamma Leonis in the end of the Sickle. Within the sweeping curve of this Sickle lies the radiant point of the Leonid meteor shower that occurs each November. There will be more on meteors later in a special chapter. The star Zeta, just above Gamma, will show three companions nearby if your eye is keen and your binoculars steady. Epsilon (ε), also in the Sickle, makes a little triangle with two 7th-magnitude stars. There is no physical connection involved here.

The letter R on the chart marks the location of a well-known variable star. At regular intervals this star becomes bright enough to be seen with 7x binoculars, and I have often seen it with the naked eye. So important are variable stars and so easily observed are many of the brighter ones that I have devoted a later chapter to the systematic study of them. Numerous charts have been included in that chapter showing just where to find them in the sky with your low-power glasses.

Directly east of the Sickle, you will find a conspicuous triangle of stars, the brightest of which is Beta (β). This star is Denebola, a name meaning "The Tail of the Lion." Below this triangle, the star Tau (τ) is a fairly easy double in binoculars. Denebola and Gamma, in the Sickle, are both standard 2nd-magnitude stars, with Gamma only .15 of a magnitude brighter. It takes practice to detect magnitude differences this small, as you will see when observing variable stars.

Regulus lies almost directly on the ecliptic — the centerline of the zodiac — so sometimes Leo is a busy place, for all the planets pass through here on their trips about the Sun. If you find a bright stranger in this region that is not shown on a chart, it will doubtlessly be one of the brighter planets. The Moon also travels through Leo in its swing around the Earth, and it will sometimes occult, or hide, some of Leo's stars from your view.

Cancer

Regulus and the Sickle lie in a rather barren sector of the sky, and this makes them stand out in greater contrast so that they serve as effective beacons to guide us to a number of fainter outlying stars and constellations. For example, halfway along a line from Regulus to Pollux is the tiny constellation Cancer, the Crab. On moonlit nights, you may have trouble finding it at all with your naked eye, though your binoculars make it easy. At the spot marked M-44, you can find the open star cluster known as Praesepe, or the Beehive, with your naked eye on a sparkling, moonless night. Your binoculars will show that this is not just a hazy cloud but a swarm of tiny stars. Galileo, with his crude new telescope, was the first to learn this. He counted thirty-six stars in the group. Closely preceding the star Alpha is the dim star cluster M-67. It is difficult in 7x glasses, for it only shows as a tiny bit of haze.

Hydra

North of Cancer and the Sickle are Lyncis, the Lynx, and Leo Minor, the Lesser Lion. These are small, faint figures that have

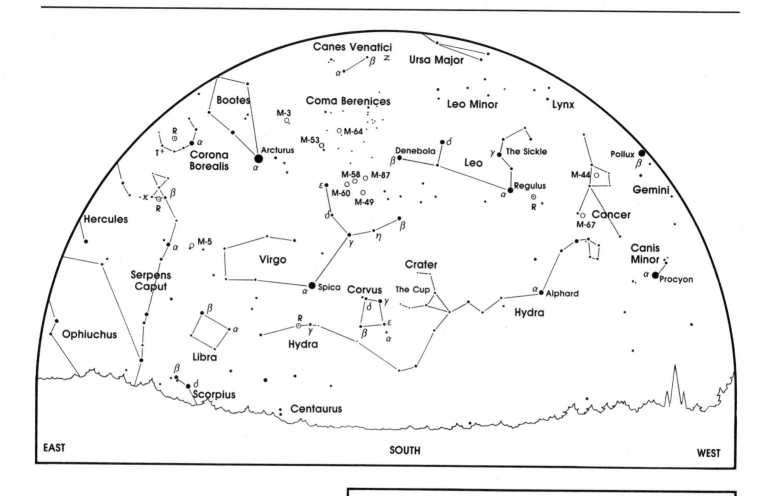

CONSTELLATIONS OF THE SOUTHERN SPRING SKY
BOOTES CANCER CANES VENATICI
COMA BERENICES CORONA BOREALIS
CRATER AND CORVUS HYDRA LEO VIRGO

THE SPRING STARS

THE SOUTHERN SPRING SKY AS SEEN ON:
MARCH 15 at 12 A.M.   MAY 1 at 9 P.M.
APRIL 1 at 11 P.M.   MAY 15 at 8 P.M.
APRIL 15 at 10 P.M.

THE SPRING STARS

nothing of importance to offer us. Directly south of Cancer, or about halfway between Regulus and Procyon, look for a compact group of five 3rd- and 4th-magnitude stars. This conspicuous figure forms the head of Hydra, the Water Snake. In total area, it is the largest of all the constellations, for it stretches its sinuous folds all across the southern springtime sky with its tail now low in the southeast. Hydra's brightest star is 2nd-magnitude Alphard, known as the Solitary One because there is no other bright star nearby. Alphard, like Betelgeuse in Orion, is classed as a red giant. Use your binoculars to bring out the redness of its fading fires. Another red star, the variable R Hydra, is charted and described in a later chapter.

Just east of Alphard and beneath Regulus is a small figure or asterism known as Sextans, the Sextant. It has no star brighter than 5th magnitude and has nothing of interest for us. Sextans is a modern invention designed by Hevelius in 1690 to fill in sky space not occupied by the adjacent constellations.

Crater and Corvus

Resting on the coils of Hydra directly in the south at this time are two small figures of ancient lineage. The westernmost of these is Crater, the Cup, a pretty design made up of faint stars in the outline of a cup or goblet. Here, again, binoculars will help to bring out its stars. To the east of Crater, Corvus, the Crow, is brighter and of much more interest, though there is little about this celestial crow to remind you of the *Corvus brachyrhynchos* of your bird guide. My old atlases show this raven facing westward and pecking at the coils of Hydra on which it stands. The stars Delta and Gamma Corvi are in the shoulders and Alpha is in the beak. Delta is a fairly easy double.

Virgo

Like the Big Dipper, Corvus has its pointer stars. The two stars in its shoulders point to a nearby bright star, 1st-magnitude Spica in the constellation Virgo, the Virgin. Like Leo, this large constellation lies in the zodiac, and therefore the Moon and planets all pass through here. So, too, does the Sun, for just west of the star Eta is the location where it stands when the days and nights are equal at the beginning of the autumn season.

Within the rough square formed by Epsilon, Gamma, and Beta

in Virgo and Denebola in Leo lies the so-called Realm of the Galaxies. Here a telescope reveals a profusion of galaxies that, because of their great distance from us, are mostly too faint for our low-power glasses, although you can catch glimpses of a few, such as M-49, M-58, M-60, and M-87.

Coma Berenices

Just to the north of the Realm of the Galaxies, which we saw fenced in by Denebola and three of Virgo's stars, is the faint constellation Coma Berenices, or Berenice's Hair. It is a widely scattered cluster of faint stars that should be looked for on a clear, moonless night. You will make out perhaps a dozen individual stars with your naked eye, while binoculars — preferably low power and wide field — will add many more. This whole region is actually a continuation of the Realm, for photographs of this area show that it, like Virgo, is strewn with thousands of galaxies, most of which are millions of light-years distant.

In my six-inch telescope, I have often seen several of these faint galaxies in a single field of view. When comet-hunting, I always carefully avoid this Comae-Virgo-Leo region because of the abundance of these objects. So many of them look just like faint comets that to check them out for possible movement would take too much of my observing time.

In legendary lore, Berenice was the Queen of ancient Egypt. When her husband the king left to battle the Assyrians, she vowed to dedicate her hair to the goddess of beauty in the hope that this act of sacrifice might assure his safe return. Accordingly, her tresses were duly shorn and deposited in the Temple of Venus, but soon they disappeared and could not be found. The King came home from battle victorious and unharmed but, understandably, became quite angry to note that Berenice's hair was gone — not only from her once-queenly head, but also from the temple. Fortunately, the court astronomer was both a stargazer and a diplomat. To console the distraught royal couple for their loss, he took them outside on a clear, dark night and pointed out to them the dimly lit area of sky where the missing locks had been placed among the stars by Jupiter.

With your binoculars you still can see that hair today, in scattered curls and ringlets, all across the constellation that was named in honor of the Queen. To some, of course, those stolen locks look more like curving lines of twinkling stars.

Near 4th-magnitude Alpha, your binoculars should easily locate the globular cluster M-53, while about four degrees to the northwest of Alpha near a 5th-magnitude star you will find the small galaxy, M-64.

Canes Venatici

Directly north of Coma Berenices is a tiny, modern star figure known as Canes Venatici, the Hunting Dogs. You will find this dim constellation as three stars creating an obtuse triangle just beneath the bend in the handle of the Big Dipper. The brightest of these is 3rd-magnitude Alpha or Cor Caroli. About three degrees south of the end star in the Dipper's handle, a little patient searching may reveal the galaxy M-51. This is the famous Whirlpool Nebula, the first galaxy observed to be spiral in form. In binoculars, it will only show as the frailest bit of haze. The globular cluster M-3, beside the southern border of the constellation, is much brighter, however. I can always locate it with the naked eye, and low-power glasses give it a bright nebulous appearance but will not resolve its stars into separate points.

Bootes

A line drawn from Beta through Alpha in Canes Venatici leads the eye almost directly to Arcturus, the brightest star north of the sky's equator. Of the fifteen 1st-magnitude stars visible in our latitude, only Sirius, the winter star, is brighter. Another way to find Arcturus is simply to follow the curving arc of the Dipper's handle to the first bright star, which is Arcturus. Continue this same curve for an equal distance and you next arrive at Spica in Virgo, which Corvus, the Crow, has already pointed out.

Arcturus stands at the southern tip of a kite-shaped figure called Bootes, the Herdsman. This brilliant yellow star — more than a hundred times brighter than our Sun — is classed as a red giant, though it is not as far advanced in its evolution as is ancient Betelgeuse.

Arcturus moves through space at a more rapid rate than any of the bright stars that we see. This speed is estimated at about 200 miles per second across our line of sight. If you have ever had difficulty in keeping a hummingbird flying at 55 miles per hour or a duck hawk at 75 miles per hour in the field of your binoculars, a speed of 200 miles per *second* may be something of a shock. You may feel that you should grab those glasses quickly and dash outside to get a look at old Arcturus before he leaves the sky completely.

But you can take your time on this, for Arcturus has been bowling along at that rate for millions of years. In the last two thousand years, it has only moved across our line of sight about one degree, or twice the apparent diameter of a Full Moon. Arcturus is 40 light-years distant from the Earth. This was dramatized in 1932 at the opening of the Century of Progress Exposition in Chicago by utilizing the light from the star to activate a switch that turned on all the exposition lights. Arcturus was given this stellar role because the light that did this had started toward the Earth 40 years before — at the time of the Columbian Exposition in 1892.

Corona Borealis

Closely following Bootes, the Herdsman, in his march across the skies, you will find the small but beautiful constellation Corona Borealis, the Northern Crown. The 2nd-magnitude star Alphecca (α), the so-called Pearl of the Crown, is the brightest star of the entire figure. Within the curve of the horseshoe-shaped outline of the Crown is the important variable star marked R.

Just outside the circlet of stars, a cross marks the location of T, a new star that rose to 2nd magnitude in 1866 then gradually faded back to 9th magnitude. Eighty years later, in 1946, it suddenly rose to 3rd magnitude and is now somewhat restless at 10th magnitude.

Here, in this tiny constellation are two important stars, the variable R and the recurrent nova T, only about four degrees apart. You should check these two locations each clear night; to help you, I have included charts and fuller details in later chapters.

CHAPTER 7
The Summer Stars

Facing North

For either the birdwatcher or the stargazer, the summer season is a busy time of year, and when you elect to follow both of these pursuits with diligence, it sometimes may seem like you are emulating Hercules and his twelve legendary labors.

The springtime flood of returned migrants has ceased the chore of nest-building and now family-raising has become the order of the day. Every year at this season, I often drive out to our old farm just to see and hear two birds in particular that have always seemed so typical of summertime and of the farm days that I knew there as a boy. I still can find them there, the meadowlark and the bobolink, and hear the flutelike whistle of the one and the joyous bubbling of the other. But, sadly, progress has not dealt kindly with these two, for both are lovers of the pasture and the meadow and these, alas, are seldom found today on modern grain farms.

Neither can the deep, soft velvet of the summer nights that I once knew so well be found on the farm today. It seems that every farmhouse now must have a naked floodlight standing guard from dusk to dawn beside it, and the skyglow of surrounding towns flares a little higher every year. The pristine sparkle of those unpolluted skies may be forever gone, but with the greater penetration of our augmented eyes, the nights are still a storehouse filled with rare enchantments. You'll find many of these treasures in the summer skies, both north and south.

Gone completely from our balmy nights is the glitter of the winter stars. Gone are the giant striding figures of Orion and the charging Bull. The bright Sickle of our springtime skies is also missing, and speeding Arcturus now is slipping toward the west. Gone, too, for the season are the frigid nights, the frosted lenses, the heavy clothing, and the mittened hands. In place of winter's rigors and the sometimes gusty nights of spring, the summer skies now offer you and your binoculars complete observing comfort and relaxation as you survey, prone upon the ground or steadied by a chair, a horizon-to-horizon floodtide of celestial sights that will often overflow the too-short hours of the summer nights.

Draco

In the north, high above Polaris and almost encircling the Little Dipper, the ancient constellation Draco, the Dragon, holds its four-sided head almost directly above you. Some say that this figure really does look like its namesake, but since I've never seen a dragon, I find instead a giant measuring-worm, its back looped high above it and its head reared back as it contentedly munches on some juicy greenery from my garden. The four stars of its head are of four different magnitudes: 2.5, 3, 4, and 5; therefore, they, like the Little Dipper stars, furnish us a scale of comparison for estimating other stars. Look very closely at Nu (ν), the faint star of the Head. Keen eyesight can split this star into two components, both 5th magnitude and just 1' (one arc-minute) apart. This is really a binary system that is easy to find with binoculars. So distant are these two from each other that at least a million years is required for a single common revolution.

Earlier we paid passing tribute to the third star in the Dragon's tail. This is Thuban, the alpha of the constellation, though it is by no means its brightest star. It can be found about half-way between the bowl of the Little Dipper and Mizar in the handle of the Big Dipper. Thuban was the Pole Star of the sky at the time of the building of the Pyramids some 5,000 years ago, and we might wish that it could shed some light today to tell us how these stony structures were erected. One wonders, too, if the Pyramids will still be standing when Earth's slowly shifting axis swings back to Thuban once again in A.D. 21,000.

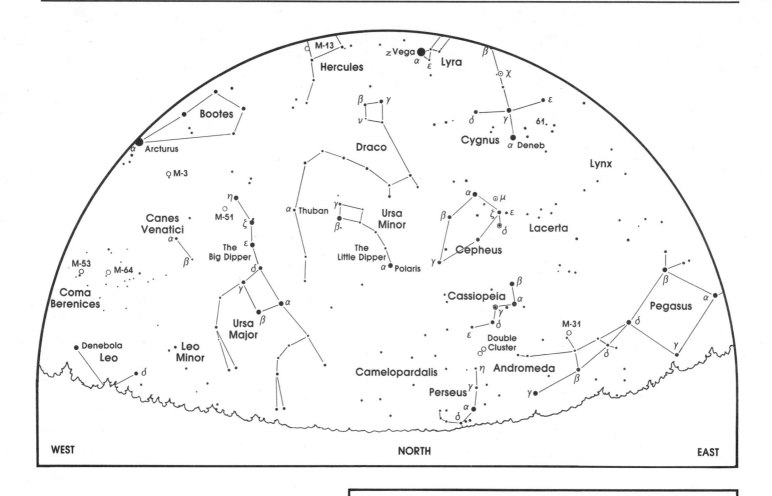

CONSTELLATIONS OF THE NORTHERN SUMMER SKY

CAMELOPARDALIS   CASSIOPEIA   CEPHEUS
DRACO   URSA MINOR

THE SUMMER STARS

THE NORTHERN SUMMER SKY AS SEEN ON:

JUNE 15 AT 12 A.M.	AUGUST 1 AT 9 P.M.
JULY 1 AT 11 P.M.	AUGUST 15 AT 8 P.M.
JULY 15 AT 10 P.M.	SEPTEMBER 1 AT 7 P.M.

Cepheus

Just east of Polaris at this time of year is an inconspicuous five-sided figure made up of 3rd- and 4th-magnitude stars that somewhat resemble the end view of a house with a steep-pitched roof. This is Cepheus, the King. For all its faintness, Cepheus has some interesting features and, as the Milky Way passes here, the entire region deserves sweeping with your binoculars. Just below the base of the "house," note the faint star Mu (μ). It is often called the Garnet Star, for it is the reddest naked-eye star to be seen from this latitude. In order to fully appreciate its color, get Mu and nearby Alpha, which is a white star, in the same field of view in your binoculars. The contrast of colors is striking. Mu is also a variable star, somewhat like Betelgeuse in Orion, and fluctuates between magnitude 3.6 and 5.8 in an irregular period.

Delta Cephei is both a double star and a variable star. Its companion is of magnitude 7.5 and its color is blue; Delta is deep yellow and varies from 3.7 to 4.6. You can watch this variation from night to night and make your own determination of its period. You'll find a chart and a more detailed account of this important star in a later chapter on variable stars. This chart also covers the variable Mu Cephei.

Camelopardalis

The large, virtually empty area extending from Polaris down to the northern horizon is occupied by the constellation Camelopardalis, the Giraffe. This must be one of the dullest of all the constellations, for it has nothing brighter than 4th magnitude and, at this time of year, you will have trouble even finding these stars, which may be just as well, for there is nothing here of interest for us.

This figure was introduced in 1690 by Hevelius who also gave us such assorted creatures of the night as the Hunting Dogs, the Lynx, the Lizard, the Little Lion, and a number of others of similar calibre that we will discuss later. Wide open spaces must have appealed to old Hevelius, for his crowning achievement was the building of a 150-foot-long telescope that had an open tube!

Facing South

Arcturus, one of the bright stars that you watched in spring, is still high up in the western sky as you face the south to begin your exploration of this inviting sector of the sky. Right beside the kite shape of Bootes, you will recognize the little circlet of the Northern Crown. With your binoculars, see if the variable R Corona Borealis is at its normal maximum brightness of about magnitude 6.0. This will only take a couple of seconds and should be a nightly ritual. Sooner or later, this strange star is certain to take a temporary leave of absence from its place within the Crown. In this same field of view also look for the recurrent nova T Corona Borealis. It is almost constant at 10th magnitude and thus a bit beyond the range of 7x glasses — but you never know when it will rise again.

Serpens and Ophiuchus

This is a mixed-up pair of constellations. Here is the giant figure of a man trying to hold a writhing serpent. (Hevelius might well have used his talents here to get this constellation straightened out.) As it appears in my old atlas, Ophiuchus occupies the center of the stage with the Serpent's head (Serpens Caput) to the west and its tail (Serpens Cauda) to the east. Ophiuchus should have been in the zodiac, because the Sun, Moon, and planets all spend much more time here than in Scorpius just beneath it.

Note the five-starred figure that forms the head of Serpens. With your binoculars locate the field of the variable R Serpentis. About once each year it rises to 5th magnitude and can be watched for several weeks before it sinks too low for 7x glasses. Beta, the brightest star in the Serpent's Head, lies near a brilliant field of stars. The globular cluster M-5 can be spotted with the naked eye about five degrees (the distance between the Pointer Stars) to the southwest of Alpha and right beside a 5th-magnitude star. Near the end of the Serpent's tail, the star Theta is a fine pair of 4th-magnitude stars. About four degrees west of it you will find a beautiful open cluster.

In the past Ophiuchus has been noted for the many novae, or new stars, that have appeared within its borders. At the spot marked 1604 on the sky chart, Kepler's Star blazed out in that year and quickly became brighter than the planet Jupiter. A few months later

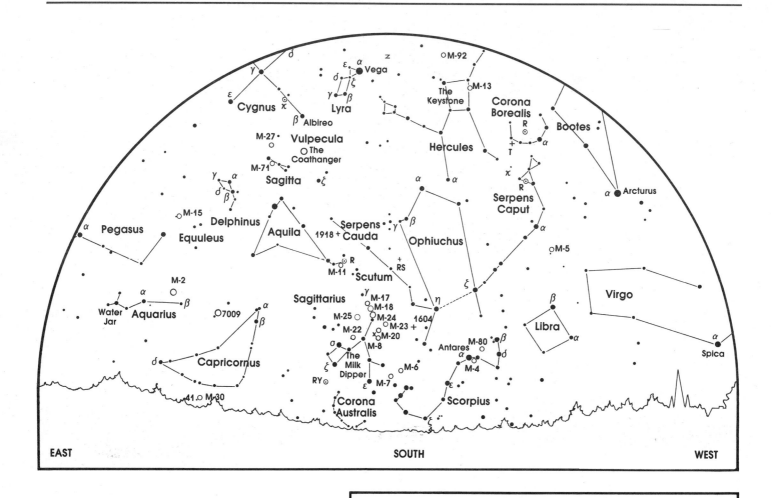

CONSTELLATIONS OF THE SOUTHERN SUMMER SKY

AQUILA CAPRICORNUS CYGNUS DELPHINUS
HERCULES LIBRA LYRA SAGITTA SAGITTARIUS
SCUTUM SERPENS AND OPHIUCHUS SCORPIUS
VULPECULA

THE SUMMER STARS

THE SOUTHERN SUMMER SKY AS SEEN ON:

JUNE 15 AT 12 A.M.	AUGUST 1 AT 9 P.M.
JULY 1 AT 11 P.M.	AUGUST 15 AT 8 P.M.
JULY 15 AT 10 P.M.	SEPTEMBER 1 AT 7 P.M.

it had faded from naked-eye visibility. RS marks the location of a recurrent, or repeating, nova. In 1898 it suddenly rose to 4th magnitude and has done the same in 1933, 1958, and 1967. It is presently a restless 11th-magnitude star that could rise again at any time. On the key chart, the location of RS is included so that you can watch for another outburst. In 1933 this star brightened about one magnitude per hour.

The Milky Way is spectacular in southern Ophiuchus, and sweeping this region with binoculars will show you many bright patches or star clouds.

Hercules

High in the sky and not far from your summer zenith is the constellation Hercules, the Kneeler. This is a difficult group, but if you have already found Draco, the circlet of Corona, the Head of the Serpent, and Ophiuchus, you have it more than half-surrounded and should have no trouble with it. Note the four-sided figure that I have marked The Keystone. The little circle near the western edge of it marks M-13, the finest globular star cluster that can be seen from our latitude. It is faintly visible to the naked eye, and your binoculars will, of course, greatly improve the image. M-13's hundred-thousand stars still merge into a hazy ball, for they are more than 30,000 light-years from Earth. Just north of the Keystone your glasses will pick up a somewhat fainter and more condensed cluster, M-92. The star Kappa (κ) is a double that may be within your powers. It can be found a little east of the Serpent's Head.

Alpha is a red giant star — one of the largest known. It is also a variable star that fluctuates in an irregular manner between 3rd and 4th magnitude in a period of about three months.

Scorpius

Beneath the feet of Ophiuchus and not far above your southern horizon you can find the bright red star Antares near mid-stream in the Milky Way. This star marks the heart of Scorpius, the Scorpion, and, for a change, this is a realistic figure. It actually does look like a scorpion with its long curving tail trailing along behind it. Like Alpha Hercules, Antares is a red giant. This modern classification is an apt one for it is the reddest of the 1st-magnitude stars and its diameter is far greater than that of the orbit of the Earth. The ecliptic, the center line of the zodiac, passes just about four degrees north of Antares, so the Moon and planets travel through here in their circuits of the sky. About every other year the red planet Mars makes a visit to Scorpius. So alike in color and in brightness are star and planet that at first glance only someone who knows the background stars can be certain which is Mars and which is Antares.

Scattered all about here are sparkling sights for your binoculars. You should always use them on a colored star like Antares, for the brighter you can make the image of a star, the truer and more brilliant will its color appear. Halfway between Antares and Beta is the rather faint globular cluster M-80. Another even better globular cluster, M-4, lies two degrees directly west of Antares. The star Zeta lies in a spectacular field of bright companions, including a small misty cluster of fainter stars. In the tail of the Scorpion are two open clusters, M-6 and M-7; they can be seen with your naked eye and are a gorgeous sight in binoculars.

On sparkling moonless nights when there is no haze above your southern horizon, sweep all through Scorpius and lower Ophiuchus, for here are some of the richest star fields of the Milky Way. When the seeing is just right you will come upon bright star clouds and dark opaque patches that take on added splendor in binoculars. Also catch with your glasses the double stars Mu and Nu.

Libra

Just west of Scorpius is the faint constellation Libra, the Scales. I have postponed mention of this figure until we had the Scorpion to help point it out, for Libra has no star brighter than 3rd magnitude. However, this small, square group is not entirely without distinction: It is the only constellation in the zodiac that represents an inanimate subject. Its name is really out of place since the word zodiac means "Pathway of the Animals." The star Alpha is a wide and easy double in binoculars, and Beta is said to be the only naked-eye star that is greenish in color.

Sagittarius

East of Scorpius and just out of reach of its venomous sting is Sagittarius, the Archer. In mythology, this figure sometimes

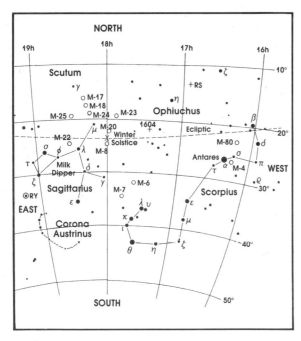

The Region of Sagittarius and Scorpius. Note the abundance of Messier nebulae and star clusters. The X on the ecliptic marks the location of the Sun at the beginning of winter.

represented a centaur — half horse and half man — with a bow and arrow aimed at Orion, who is now completely out of sight. You can see the bow formed by the curved line of the stars Lambda, Delta, and Epsilon; the tip of the arrow is Gamma. The ecliptic, and therefore also the zodiac, passes entirely across the constellation. At the beginning of winter around December 22, the Sun occupies the position on the ecliptic marked X on the key chart. From that point on, it will slowly move northeastward to the spot in Gemini where it stood at the beginning of summer.

Sagittarius has a wealth of outstanding objects that are certain to challenge your binoculars and your observing skill when you search these southern star fields on pleasant summer nights. So closely grouped are many of these points of interest that I have included a detailed chart of this region to assist you in locating them more readily. Near the western border, note M-8; this is a star cluster superimposed on a hazy nebula that is readily visible to the naked eye. Just north of it is M-20, the Trifid Nebula. It can be seen in binoculars, though it takes a telescope to bring out its three-part form. In this same region, M-17, M-22, and M-24 are all worth careful study.

In thickly populated areas such as these, explore the entire region — not just the few special objects that I have cited but rather the whole constellation — with your binoculars very slowly and carefully. Sagittarius lies in perhaps the most exciting sector of the entire sky. Here the Milky Way is at its brightest and its broadest. Here are the finest star clouds and dark nebulae to be seen from this latitude. And here the densest concentrations of stars anywhere about the skies are to be found. When you turn your glasses on Sagittarius, you are probing deeply into the very anatomy of our Galaxy, as I will explain in the chapter devoted to the Milky Way.

RY Sagittarii is an important variable star that you should locate and add to your observing list. It is the same irregular type as R Corona Borealis and, like it, is usually within easy reach of binoculars. But it, too, can fade from sight quite suddenly.

Lyra

High in the sky on summer nights is blue-white Vega in the

constellation Lyra, the Lyre. It is the brightest of the summer stars. You need no pointer stars to guide you to it for Vega's color and her five nearby companions make identification easy. Vega is a youthful star, which is fortunate, for this makes us even more certain that she will keep her appointment to be Earth's pole star in a mere 12,000 years from now.

The first star to the east of Vega is the "double-double" Epsilon. On a clear night, your naked eye can divide it. My old atlas shows it as two stars side by side, both magnitude 4.5. Binoculars will widen the gap between them but will not, as a telescope will, again divide each of these two stars into a pair. Zeta is also double in binoculars, if they are steadily held on some support. I recall that as a rank beginner in my early teens I "discovered" this double quite by accident one summer night and naively wondered if I should report my find to some observatory! Beta is a variable star ranging between magnitude 3.4 and 4.3 approximately every 13 days. The spectroscope shows that Beta is a binary system with its two components nearly in contact. Their strong mutual gravitation has pulled them into egg-shaped rather than spherical forms. You can follow these variations yourself using Gamma (magnitude 3.2) and Kappa (4.3) as comparison stars.

Cygnus

Following directly after Lyra, its neighbor to the east is Cygnus, the Swan. More familiarly known as the Northern Cross, you can easily trace out the two axes or arms of this figure. The long one lies in the Milky Way with 1st-magnitude Deneb at the northern end and 3rd-magnitude Beta or Albireo at the southern foot of the Cross, where it represents the head of the fabled swan as it flies south. Albireo is one of the finest double stars in the sky; its components shine at magnitude 3.2 and 5.4 with colors of contrasting orange and blue. It is believed to be a binary system of extremely long period.

The star marked 61 is a double of magnitudes 5.6 and 6.3. At 10.5 light-years away, it is one of our nearer neighbors and was the first star whose distance was directly measured. Omicron (o) is a wide double whose magnitudes are 4.0 and 5.0 and whose colors are yellow and blue.

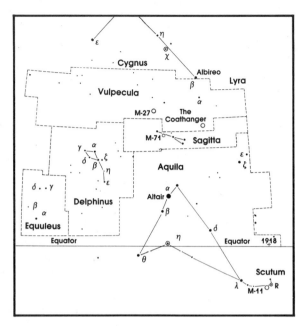

Detail of a group of smaller constellations

The northern Milky Way is at its finest in Cygnus where the region around Gamma, at the crossing of the arms, glitters with a treasury of gems for your binoculars. Find the Northern Coal Sack near Deneb on some night of near-perfect seeing; it is an irregular cloud of opaque dust that blots out the brighter stardust of the Milky Way behind it. The stars that you see scattered all about lie between you and the Coal Sack.

Use your glasses to search for the variable star marked with the letter Chi (χ) in the long axis of the Cross. You may be able to follow it for about three months of its cycle, as it sometimes brightens to 4th magnitude.

Aquila

If you are a birdwatcher your heart should now feel all aglow, for I have just pointed out to you Cygnus, the Swan, and now comes Aquila, the Eagle. Altair, the bright star of this celestial raptor, has as both habitat and flyway the mainstream of the Milky Way, and as in a proper bird book, it has a star on either side to serve as field marks. Just a little way below this trio, the star Eta is an interesting variable star that can be followed throughout its entire range. In Chapter 10, I give comparison star magnitudes for this Cepheid-type variable. Watch it carefully and make your own determination of its period. Note where a bright nova suddenly appeared in 1918. For a night or two it equalled Sirius, the brightest of the stars. Today its light is steady at magnitude 11.

Altair is the third point of the celestial figure called the Summer Triangle. Together with Vega and Deneb to the north, this bright 1st-magnitude landmark of the skies will be with us all through our summer and autumn evenings. Because Altair is farther south than either Vega or Deneb, it is the last of the three to arrive on the scene and also the first to leave. Watch for it near midnight in the east on a mid-April evening and bid it *au revoir* at early dark in January.

Scutum

This tiny figure is another invention of Hevelius. Scutum, the Shield, lies in an area of the Milky Way that is sprinkled with luminous star clouds, the brightest being the open star cluster M-11. It can easily be detected with the naked eye and makes a fine appearance in binoculars. At a distance of 4,300 light-years, it belongs to our own galactic system. This entire region is well worth exploring with your glasses. The variable R Scuti is an important star because of its irregular behavior. It is always within the powers of your binoculars, and you will find it interesting to keep a record of its changes. It also shows the variable Eta Aquilae.

Sagitta

Halfway between Altair and Albireo at the foot of the Northern Cross you will find Sagitta, the Arrow, in a fine region of the Milky Way. The rather faint cluster M-71 lies on the shaft of this realistic-looking arrow formed by the constellation's four brightest stars. Ardent birdwatchers may prefer to disregard this figure, for it represents the arrow that Hercules has just shot at the two birds, Cygnus and Aquila. I'm sure he missed them both, for the arrow, as we see it, is passing harmlessly between our feathered friends of the Milky flyway.

Vulpecula

Lying between Cygnus to the north and Sagitta to the south is a long, narrow area dimly lit by a few faint stars. Once again Hevelius has been here before us. It seems that he had a creative urge, for here he has taken a scattered handful of 5th-magnitude stars and made them into a long, slender fox that carries in his jaws a captured goose — the third bird in this congested aviary of the skies. This constellation, Vulpecula, the Fox, is noteworthy as the home of M-27, the famous Dumbbell Nebula (only a misty spot in our glasses), and for the number of novae that have been discovered within its borders.

Just north of the Vulpecula-Sagitta border you will find a remarkable assemblage of 6th- and 7th-magnitude stars. This is popularly known as "The Coathanger" and, even though it appears upside down in your binoculars, you can see at once its likeness to that utilitarian contrivance. This resemblance was briefly disturbed in 1976 when a 6th-magnitude nova suddenly shone out beside the long bar of the hanger.

Delphinus

Directly east of Aquila your eye is drawn to the bright, compact figure of Delphinus, the Dolphin. This is a diamond-shaped group of 3rd- and 4th-magnitude stars that, for no apparent reason, is often known as Job's Coffin. The star Gamma is a double of magnitudes 4.5 and 5.5 at a distance of 12 arcseconds (12"), which I find too close to divide with 7x.

Delphinus is the brightest of a confusing group of little constellations without a single star of any note. The chart of this area shows the boundaries of Delphinus, Sagitta, and Vulpecula as related to Cygnus and Aquila. I have also added Equuleus, the Little Horse, which was tethered in this barren spot by Ptolemy of Alexandria some 2,000 years ago.

Capricornus

A line pointing southeast through Altair and his two companions points directly to Capricornus, the Sea-Goat. To the eye, this figure resembles the outline of an inverted cocked hat of Revolutionary War vintage. It is a constellation for late summer nights. Capricornus never rises high above the southern horizon, so only in the best of seeing will its stars be sharp and steady.

If you found Mizar and Alcor in the Big Dipper to be no challenge for the naked eye, now try Alpha in Capricornus. A clear night will show it to be a double of magnitudes 3.6 and 4.3 lying side by side. With steadied binoculars you may be able to again divide each of these two stars, though the companions are 9th magnitude and difficult. The two bright stars of Alpha are not physically connected by gravitation; their independent motions through space are slowly increasing their distance from each other. Not until the 17th century could these stars be separated by the naked eye. Beta is an easy double. The brighter star is orange and the fainter companion is blue in your binoculars. The colors of the pair remind me of Albireo in Cygnus, though here the distance between the stars is much greater. A third easy double in your glasses is the faint star Omicron. With your binoculars locate the 5th-magnitude star marked 41. Right beside it to the west is the faint globular cluster M-30.

Capricornus lies in the zodiac, so all the major planets pass through here in their trips around the Sun. One of them, in fact, was captured here. The planet Neptune was discovered near Delta in 1846.

CHAPTER 8
The Autumn Stars

Facing North

In the western sector of the sky at the beginning of this season, the three stars of the Summer Triangle — Vega, Deneb, and Altair — are still in good position for observing while, lower in the twilight, Arcturus leaves our evening skies in mid-September. Above the skyline in the west, you still can watch the strange variable R Coronae for any sudden changes and, if you are vigilant, you may also find its brother, irregular RY Sagittarii, low in the southwestern sky.

The Milky Way, which in spring lay so low along the northern horizon that it sometimes drowned in haze, now stretches high across the northern sky from east to west and can be observed with ease. With the advent of the longer nights of autumn, it is Ursa Major, the Great Bear, that now becomes entangled in the maze of mist and treetops in the north.

Unless the night is one of sparkling clearness, you may have trouble making out all seven stars of the Big Dipper. Just a glimpse of them, however, will serve to guide you on. A line drawn by your eye from the middle of the Dipper figure through Polaris then onward an equal distance will bring you to a compact group of five bright stars known as Cassiopeia.

Cassiopeia

According to legend, Cassiopeia was the queen of Ethiopia and the wife of Cepheus who, quite dull by comparison, stands beside her to the west. At this time of year, Cassiopeia forms a letter "M" in the sky, though as time goes by and season follows season, this

"M" slowly changes to next spring's letter "W." But it will still be just across Polaris from the Dipper.

Alpha is a double star that requires a sharp eye and a steady glass to separate, since the small blue companion is only 9th magnitude and is virtually flooded by the scintillation from the brighter star. The distance between them is some 60" (60 seconds), which would be far enough apart to distinguish the two stars in binoculars if they were of equal brightness. Gamma, in the center of the "M," is a remarkable variable star of the most erratic type. The cross marked 1572 shows the location, in that year, of the brightest nova ever recorded. It became brighter than the planet Venus and, for a time, could be seen in full daylight. It is known as Tycho's Star because the great astronomer Tycho Brahe kept an accurate record of its changes for seventeen months until it became invisible to the naked eye.

Cassiopeia abounds in scattered groups and clusters of stars of which M-52 and 663 are conspicuous. The entire region provides interesting sweeping since the Milky Way passes through here.

Perseus

Moving eastward from Cassiopeia, you will at once encounter the martial figure of Perseus, the Hero. This bright and beautiful constellation doesn't resemble anything of a familiar shape. It is simply an assortment of curving lines of stars that seem to come together at the star Delta, near the center of the arrangement. The region that lies beneath Alpha is one of surprises and delights. Sweep your binoculars carefully through this sector to discover for yourself the many fields of visual charm and wonder that await you here.

Make believe that you are Galileo with the world's first celestial spyglass. He had no chart or guidebook to the skies, but when he turned his homemade scope on a hazy speck of light that he could glimpse in northern Perseus, it was transformed into a maze of tiny, twinkling stars. This famous Double Cluster, which bears the designation h-chi (h-χ), can easily be seen on moonless nights making a triangle with the Delta and Epsilon stars of Cassiopeia.

Look at Beta Persei. This is Algol, the Demon Star, and the constellation's most noted attraction. This star is the prototype of a whole class of variable stars. If you watch Algol faithfully from night to night, you will find some point in your vigil when its light suddenly begins to fade. It drops steadily for some 3½ hours to magnitude 3.4 and remains at that level for 18 minutes; it then takes another 3½ hours to get back to its original magnitude of 2.2 — nearly as bright as Polaris. Algol remains bright for 2 days 20 hours 48 minutes and 55 seconds, and then the entire cycle repeats.

Strictly speaking, Algol is not a variable star at all, although the light that we receive from it certainly does vary. It all depends on your viewpoint. If the Earth were located in some other part of the universe, Algol's light would burn for us as constant as the Sun's. It is, instead, an eclipsing binary, a 2nd-magnitude star with a much fainter and somewhat larger companion star nearby. Together the two stars revolve about their common center of gravity in the above named period. However, the plane of their rotation is tilted somewhat to the Earth so that, from where we stand, neither of the stars ever totally eclipses the other.

When these facts are known, the Demon Star is robbed of its dark mystery. Having watched one eclipse and timed it, you can, knowing its precise period, make your own schedule of future eclipses. You can, in fact, become a latter-day Columbus and tell the local natives exactly when a certain Demon Star will begin to lose its light.

Just above Algol, the open cluster M-34 is easily found; it has a nebulous appearance in binoculars. Near the Double Cluster lies the radiant point from which the Perseid meteors seem to appear each year on or about August 12. (More on meteors in a later chapter.) In February 1901, a nova about as bright as Vega or Arcturus suddenly appeared at the spot marked with a cross. Today, it is still a restless star of magnitude 13.

Auriga

Continuing eastward along the broad pathway that we found so studded with the stars of Cassiopeia and Perseus, we come to the pentagon formed by the five brighter stars of Auriga, the Charioteer. Capella, a star as bright as either Vega or Arcturus, is the undisputed ruler of this region and the most northerly of all the 1st-magnitude

THE NORTHERN AUTUMN SKY AS SEEN ON:

SEPTEMBER 15 AT 12 A.M.   NOVEMBER 1 AT 9 P.M.
OCTOBER 1 AT 11 P.M.   NOVEMBER 15 AT 8 P.M.
OCTOBER 15 AT 10 P.M.   DECEMBER 1 AT 7 P.M.

stars. You may note that Auriga seems to have two stars named Beta — one to the north, the other in the south. Actually the southern Beta marks the tip of one of the horns of Taurus, the Bull, though it is also usually included in the pentagon.

Near Capella, you will find a small acute triangle of 3rd- and 4th-magnitude stars often called The Kids. Epsilon, at the apex of the triangle, is a most remarkable star. Like Algol in Perseus, it is an eclipsing binary, but where the period of Algol is about 2½ days, that of Epsilon and its dimmer companion is 27 *years*! Zeta, the fainter of the stars in the base of the triangle, is still another eclipsing binary, with a period of a little more than 2½ years.

The Kids provide a good means of identifying Auriga and Capella at any time or season and at any position in the sky. In the lower portion of the constellation you will find rich sweeping for your binoculars, for the main stream of the Milky Way flows through here. You can readily discover such starry islands as the open clusters M-36, M-37, and M-38.

Facing South

This is the season of the sere and brown. This, the time of harvest is also a period of plenty: a time when rural fields display their offerings of ripening grain, and every hedgerow, vine, and roadside weed presents its largesse of berries, fruits, and seeds. Yet in the midst of all this abundance, the birds that we have seen and heard all summer seem restless and preoccupied.

Soon, as though in response to some mysterious voice calling to them from a southern marsh, a Texas offshore island, or perhaps a jungle in Brazil, they take their leave of us, unwitnessed, to wing their way along some hereditary route to keep an age-old rendezvous throughout the winter months.

> Thus in the winter stands the lonely tree,
> Nor knows what birds have vanished one by one
> Yet knows its boughs more silent than before.
> — *Edna St. Vincent Millay*

With the approach of winter, the starry birds of our summer night skies will also vanish one by one. Cygnus, the Swan, is now nightly winging westward, its head outstretched toward the horizon. Low in the southwestern sky, Aquila, the Eagle, will soon be leaving for its migration flight along a star-flecked, milky flyway. We do not mourn the seasonal passing of either birds or stars, for we know that with the coming spring they'll all be back again. And in their absence they have left us with replacements of interest and charm.

Just as the migrating bobolink and flicker bring back to our doorstop the chickadee and junco, so too the dropping in the west of the starry figures of summer brings up in the east a host of visual delights that will keep our eyes and binoculars pointed skyward through all the nights of autumn. High up in the east, you'll find a horse with wings, a maiden bound in chains, and a neighboring universe. In deep mid-skies, you'll sight a monstrous whale, and, far to the south, a lonely fish with a single bright star in its mouth.

The Southern Fish and Fomalhaut

Just a little west of due south on a mid-autumn evening, you will find 1st-magnitude Fomalhaut in the constellation Piscis Austrinus, the Southern Fish. Fomalhaut is the most southern of all the 1st-magnitude stars visible from our latitude. It is almost exactly the same brightness as Pollux, a part of Gemini, the Twins, which you saw in the winter months and which already may be rising low down in your eastern sky. There is nothing about this Southern Fish that suggests one of the finny tribe, but Fomalhaut does make one corner of a small, four-sided figure. Since it is the only bright star in this region of the sky, identification is easy. The star Beta, in the opposite corner of the quadrilateral, is a double star, though horizon haze may make its 8th-magnitude companion difficult to see in low-power binoculars.

Fomalhaut rises low in the southeastern sky at almost the same instant that Capella appears above the northeastern horizon. However, Fomalhaut is so far south that it makes its brief march across the sky and sets long before Capella reaches the meridian, its half-way mark.

Pegasus

High up in the sky directly above Fomalhaut, you will find a characteristic figure of the autumn nights — the Square of Pegasus.

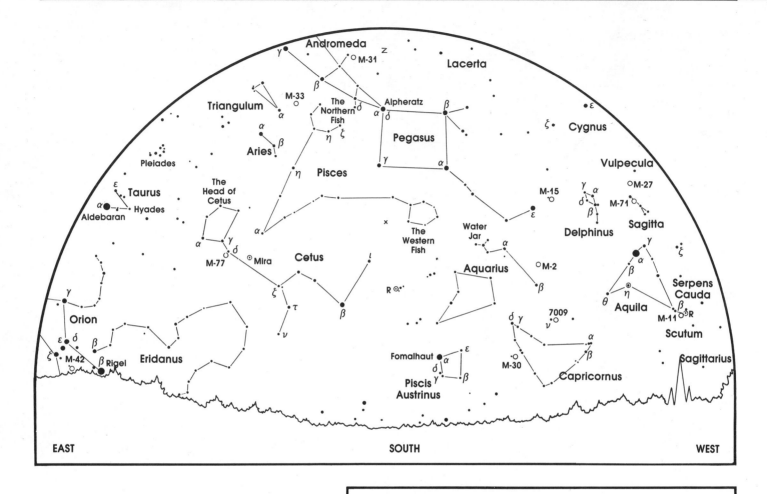

CONSTELLATIONS OF THE SOUTHERN AUTUMN SKY

AQUARIUS ARIES ANDROMEDA CETUS PEGASUS
PISCES PISCIS AUSTRINUS TRIANGULUM

THE SOUTHERN AUTUMN SKY AS SEEN ON:

SEPTEMBER 15 AT 12 A.M.   NOVEMBER 1 AT 9 P.M.
OCTOBER 1 AT 11 P.M.   NOVEMBER 15 AT 8 P.M.
OCTOBER 15 AT 10 P.M.   DECEMBER 1 AT 7 P.M.

Here are four stars of almost 2nd-magnitude brightness making a square of nearly equal sides. Note that the two stars that form the west side of the square point directly downward to Fomalhaut. The brightest of the Square stars is Alpheratz in the upper left-hand corner, though it does not really belong to Pegasus but to Andromeda, as we shall see later. This is another case of star-borrowing, such as was done with Beta from the horn of Taurus in order to complete the pentagon of Auriga.

Pegasus, the Winged Horse, lies in a sparsely populated region of sky, and for this reason not more than a figurative handful of stars can be seen with the naked eye within the large area enclosed by the Square. The globular cluster M-15 is easily seen with binoculars, though the individual stars that compose it cannot be resolved without a telescope.

Aquarius

Just east of Capricornus and north of Fomalhaut lies the large, dull, shapeless constellation Aquarius, the Water-Bearer. The fact that it is a zodiacal constellation gives it some added importance. It was near the star Delta that an unknown planet was once seen but not recognized. In 1756, Tobias Mayer almost discovered Uranus here, but it was moving so slowly that he recorded it as a fixed star. It was not until it had crept eastward into Gemini twenty-five years later that William Herschel found it again. And he at first mistook it for a comet.

Note the four-starred figure called the Water Jar and, to the east, the streams of small stars that represent the water pouring from its mouth. The globular cluster M-2 is a good subject for binoculars, though you must look sharp, because it is a compact cluster that can be mistaken for a star. On the best of nights I can faintly see it with the naked eye. With your glasses sweep the entire region. The peculiar variable star R Aquarii, unfortunately, is generally too faint for our binoculars. About a degree and a half west of the star Nu lies the Saturn Nebula. It is a gaseous nebula, which your binoculars may show as just a speck of haze. On the key chart it is numbered 7009.

Pisces

Two faint streams of stars border the Square of Pegasus on the south and east. With the naked eye only the Circlet, which represents the open mouth of the Western Fish, is bright enough to be easily made out. With binoculars both the Western and the Northern Fish can readily be traced, though in the Northern Fish Eta and Zeta lie in Andromeda's territory. At the spot marked "X", the Sun crosses the equator heading northward on the first day of spring. The ecliptic, the equator, and a line passing through both north and south celestial poles all intersect at this point. This north-south line is the zero hour circle, and all star positions are measured east from it. It is the Greenwich Meridian of the sky.

In my old sky atlas, these two fish are tied together with a long ribbon attached to their tails. Two winding streams of faint stars represent this ribbon, which loops around Alpha, the constellation's brightest star.

Cetus

If you follow the faint star stream that ties the Western Fish to Alpha, it will lead you almost to the Head of Cetus, the Whale. Appropriately, this is one of the largest of all constellations in area, though there is nothing about it that will attract your casual attention. Beta is its brightest star at 2nd magnitude. The head of the monster is marked with 3rd- and 4th-magnitude stars in the shape of a small pentagon, which you will soon learn to recognize.

Cetus's chief claim to fame is a star that you may not find at all on your first search. It is Mira, or Omicron, the first variable star ever discovered. It was found in 1596 by the astronomer David Fabricius who noted the star as being 3rd magnitude in August of that year. By autumn it had disappeared.

The star marked 66 in the field near Mira is a close double in binoculars. You might say that 66 is northwest of Mira, which it is, but in giving sky directions it is less confusing to use the term north-preceding since 66 precedes Mira in its path across the sky. (If 66 were southeast of Mira, it would be referred to as south-following.) Alpha is a wide double in binoculars; its magnitudes are 2.5 and 5.5.

Closely south-following the star Delta, look for the faint spiral galaxy M-77. It is an elongated, misty spot whose dim light has travelled perhaps 40 million years to reach us.

Andromeda

Finding this constellation is a simple matter. The north-following star in the Square of Pegasus is also the Alpha star of Andromeda. From this starting point the constellation figure sweeps to the left in two long, curving arcs. The lower arc contains three 2nd-magnitude stars that point almost directly to 2nd-magnitude Alpha, or Mirfak, in Perseus.

The outstanding attraction in Andromeda is the Great Galaxy, M-31. You can easily find it with your naked eye on a good night, and binoculars make it a brighter, hazier patch of light that over the years has often been reported as a comet. At a distance of 2 million light-years, it is the most remote object the unaided human eye can see. In 1885, a 6th-magnitude nova suddenly appeared in M-31. To seem so bright at such an immense distance, it may well have been the most luminous star ever seen. Man had not yet appeared here on the Earth when that star's explosive light started on its way toward us!

Several of autumn's best constellations are prominent in mythology's *Who's Who in the Skies*. Two of these are King Cepheus and his wife Queen Cassiopeia, both of whom, you'll remember, we have already met in our exploration of the northern skies. When the Queen boasted that she was fairer than the sea nymphs, the wrath of Neptune, the god of the seas, was aroused. He decreed that Andromeda, the beauteous daughter of the royal pair, must be sacrificed. Accordingly, she was chained to a rock beside the sea and left to be devoured by Cetus, the terrible monster of the deep.

Fortunately, however, our hero Perseus, on learning of these goings-on, mounted the winged horse Pegasus and dashed to the rescue of the maiden. Mythology further tells us that Andromeda rewarded Perseus for his valor by becoming his wife. They must have lived happily ever after for, as you can see, they still are sparkling side by side. And so this little drama of the skies ends on a happy note, a fitting final curtain for a truly all-star cast.

Triangulum

Beneath a line connecting the Beta and Gamma stars of Andromeda you will find a long acute triangle. Appropriately, this small constellation is called Triangulum, the Triangle. Its stars are of 3rd- and 4th-magnitude. With your glasses look for M-33 to the right of the sharp point of the triangle. Like M-31 in Andromeda, this is also a galaxy.

I have often glimpsed M-33 with the naked eye but only when it was high up in the sky and thus well above the haze of the horizon. You must have a clear, dark night and a location free from all interfering lights. With binoculars you will find it much fainter than M-31. Though M-33 is at its visual best in low-power glasses, you will get no hint of its marvelous spiral pinwheel structure that is so striking in long-exposure photographs.

Ceres, the largest of the asteroids and the first to be discovered, was found in Triangulum on January 1, 1801 — the first night of the new century.

Aries

Here is another triangle, though this one is obtuse in form. It lies between Triangulum and the Head of Cetus where its only conspicuous figure is made up of three stars of 2nd, 3rd, and 4th magnitudes. Aries, the Ram, has little of observational interest for us unless it has planetary guests. Being a zodiacal constellation, the Moon and planets pass through here to get from Pisces into Taurus. Two thousand years ago, the point where the Sun crossed the equator on the first day of spring lay in Aries, though now, as we have seen, it lies in Pisces. Aries is still the first sign in the zodiac of the astrologer and the old-time fortune teller even though the Sun, because of that peculiar wobble in our polar axis, has long since cancelled its appointment with the Ram and won't arrive here for another month.

The stars and constellations of the four seasons have now crossed our skies — a sparkling pageant passing in review before us. In this throng you may, from time to time, have seen individuals that stirred your imagination, challenged your credulity, or taxed the modest powers of your binoculars. Still others in this bright parade may have quite eluded you, perhaps because of clouds or moonlight or for other reasons. If so, they will all be back, and you can try again, for this is an annual procession that never ends.

    I have remarked before that as each season in the skies gives way to the next, the advance scouts of the approaching company are already well in sight. Even as you watched small Aries on autumn nights, you must have seen, just a little to the east, a tiny, misty group of stars, the Pleiades, that should now be old acquaintances for they danced their way all across our winter skies. Just as the chilly days of fall bring returning juncos and redpolls to our feeders, the crisp autumn nights bring to the eastern sky the baleful eye of Taurus. When, on a clear night in November, I hear the five notes of a Great Horned Owl coming from the shelter of our pines, I also see the giant figure of Orion standing just above the treetops as though eager to begin his nightly march across our winter skies.

CHAPTER 9
The Milky Way

In the sky map drawings you saw the staggered path of the Milky Way across the skies of the seasons of the year. In the winter skies it is a rather narrow band that rises upright from the northern horizon, passes nearly overhead, then, like a river lighted by bright beacons flanking either shore, flows southward until it drops behind the Earth just below the great star Sirius.

In the evening skies of spring, the Milky Way shows only in the north, though you must have a sparkling night and a clean sweep all along your northern horizon if you want to find it. Rely on your binoculars to brighten up this strip of sky, for the "milkiness" is overwhelmed by mist. The course of this dim trail is best marked by the constellations through which it passes. From west to east, Hiawatha's "pathway of the ghosts" skirts the northern skyline from southern Auriga through Perseus, Cassiopeia, Cepheus, and the Northern Cross.

In summer, and particularly in the month of August, the Milky Way is at its best. This is the season of the year when, on a sparkling, moonless night, you should arm yourself with your binoculars and make a special expedition to some favorite country spot that has a clear horizon all around and no distracting lights. There you will see the Milky Way in all its glory.

Many years ago I was the guest, one Texas night, of an observing group that regularly drove some forty miles to a lonely desert hilltop to carry out their program of estimating variables, plotting meteor trails, and photographing special deep-sky objects. While my companions were setting up their instruments, I walked alone to a completely isolated rise a couple of hundred yards away and looked about me. From that arid, treeless spot I could just make out the far horizon as a dim, unbroken circle where the darkness of the Earth gave way to sparkling stars. Below and all around there was only empty blackness, but across the sky above me the Milky Way, brighter by far than I had ever seen it, was like a long, slack line supported in the zenith by the Northern Cross and hung from end to end with all the trailing, filmy garments of the night that might just have been washed and hung up high above the Earth to dry.

Soon my friends came searching for me and I returned with them to their observing site. But the eerie setting — the utter silence of that lonely desert hill, the velvet blackness of the night and, over all, the bright course of the Milky Way with its pools of murky shadow and stepping-stones of starclouds — made that night an occasion that I still hope will sometime come again.

Obviously, such lucid nights can never happen in most people's environments. The world is too much with most of us. There are just too many people, too many lights and trees and buildings all about us to see such unrestricted skies. But we can try for something like it when we travel or are on vacation to some unspoiled and uncrowded area. You will doubtless never spot a Bohemian Waxwing or a Kirtland's Warbler in your own backyard, but they do exist somewhere and, as with celestial rarities, they can be seen if you go to meet them.

Let's assume that you have found a good location for observing and that it's a clear, moonless summer night. Depending on the evening hour, some of the brighter stars of Perseus should now be showing in the low northeast. You have already swept this region with binoculars in the autumn season and will recall the brilliant streams of stars your glasses revealed near Mirfak, the constellation's brightest star. Above it lies the famous Double Cluster and the five bright stars that form the "W" of Cassiopeia. Still higher in the sky, the Milky Way flows by the foundation of the "house" of Cepheus, and right here you can pause again to note the brightness of the variable Delta Cephei and the redness of Mu, the Garnet Star. This brings you to the borders of Cygnus and the Northern Cross. Here the whole character of the Milky Way seems to change abruptly. Throughout this entire northern sector from Perseus to Cygnus, the Milky Way has been a single, rather narrow trail liberally sprinkled with a host of distinct, sparkling stars. But it hasn't been a dusty path. There has been little of the hazy or "milky" appearance that you will encounter from this point southward.

Two large, dark areas bridged by a brighter spot can be seen near Deneb at the top of the Northern Cross. This region is sometimes called the Northern Coal Sack. Halfway down the long arm of this Cross, just south of the star Gamma, you will find the beginning of the long, dark lane known as the Great Rift, which for a distance divides the Milky Way into two narrow streams. At the start, the western branch is the brighter of the two, but this soon fades and completely disappears in the region of the tail of Serpens. Meanwhile, across the Rift, the eastern branch of the divided stream steadily increases in brightness through Sagitta and Aquila, with Altair and its companion stars glowing near the center of this branch. You will find one of the brightest spots of all in the tiny constellation Scutum. This is the open cluster M-11 with its bright star in the center. Just to the west of it, you will want to pause again to jot down your estimate of the perplexing variable star R Scuti, which is always well within the reach of your binoculars.

South of Scutum, the Great Rift leads into Sagittarius. Here, like sunlit islands in a stream, are some of the broadest star clouds of the entire Milky Way. Sweep carefully over the region below the handle of the inverted Milk Dipper. M-20 and M-8 are outstanding and can be detected without binoculars. In lower Sagittarius, the Great Rift ends and the Milky Way attains its broadest width, extending well into Scorpius to include red Antares among its sparkling fields. Search the region about two degrees above Antares where 5th-magnitude Rho (ϱ) Ophiuchi is the center of a dark nebulae. These seemingly blank areas, which have been called "The Windows of Absolute Night," are not holes in the starry fabric of the Milky Way but are actually dark clouds of cosmic dust obscuring the light of the myriad stars that shine unseen behind them. The Coal Sack in Cygnus, as well as the long line of the Great Rift, are fashioned out of this same non-luminous material.

You will note, both with your naked eye and with your binoculars, that this southern section of the Milky Way, particularly in Sagittarius and Scorpius, is much brighter and has much more diffuse haze than the northern path. This is why the Milky Way of autumn is less spectacular than that of summer, for in early fall this bright, broad region is setting in the west and is lost to sight. So, while summertime and Sagittarius are still with us, let's try to grasp the real significance of this "broad and ample road whose dust is gold and pavement stars." Let's try to find our station on that dusty road and place a marker there.

Either with the eye or binoculars, it's difficult to comprehend the appearance and structure of the Milky Way. You can only see it "through a glass, darkly," for you are watching from a point that lies well *within* the Milky Way. You are on the inside looking out. But you are with noble company, for all the stars you see at night — together with the Sun and his family of nine planets, including the planet Earth — are fellow members of this same great galactic system. In a sense, we also are its prisoners, for only in imagination are we free to leave this tenuous assemblage of stars and cosmic clouds and interstellar dust to view it from outside its borders.

Scattered all about the skies in every direction are millions of other Milky Ways that are more or less like ours. You have already been introduced to some of them in our autumn skies: M-33 in Triangulum and M-31 in Andromeda were pointed out as nearby and bright examples. Some galaxies turn their faces directly toward us and others present an edge-on view. We see M-33 as though we were looking directly at its center, just as a sparrow hawk might view a clover field as he hovers above it in mid-air. This is possibly what our Milky Way Galaxy would look like to some stargazer watching us from a planet in M-33.

On page 83 is a drawing showing how our galaxy might appear in this stargazer's super-telescope. The spiral arms of this celestial pinwheel are so characteristic of this type of galaxy that they are often called spirals or spiral galaxies. M-33 has spiral arms, so has M-31, and so, too, astronomers believe, has our own galaxy.

In the other drawing we see what our Milky Way probably would look like if we could see it from an edge-on viewpoint. This, too, is an inferred representation, for there are many of these edge-on spirals in the sky. None are clearly visible in binoculars, though I have often seen them in my telescope, with fine examples in Coma Berenices and Ursa Major.

The small crossmarks in each of these two drawings show the location of our solar system within the Milky Way. Inasmuch as the Milky Way is some 100,000 light-years in diameter and about 10,000 light-years in thickness, our Sun appears to be about 30,000 light-years from the center and midway between the edges of the galactic disk.

But what proof have we that these positions are correct? With nothing but your eyes and your binoculars, you can see much of the evidence for yourself. Look again at the edge-on view and imagine that you are standing at the crossmark looking toward the bulge in the center. That bulge is the center of our galaxy, and we saw it in the summer skies when we looked at Sagittarius and remarked how broad and bright the Milky Way was in that region and how thickly the star clouds were heaped about when we swept our binoculars through there. Now, in imagination, turn around and look directly *away* from the center bulge. You are now looking at the winter Milky Way, at such regions as Taurus, Gemini, and Monoceros where, you will recall, there were no star clouds and the stars themselves were relatively few and far between compared to Sagittarius and Scorpius in our summer skies.

To verify the Sun's location near the middle of the edge-on view of our galaxy, select a time when the Milky Way is overhead and divides your sky into equal halves. Now roughly compare the number and distribution of the stars you see in these two halves. As these halves seem to be about equally populated with stars, it shows that you must be standing near the marked position and are looking toward the regions of the north and south galactic poles. The dark lanes that you see crossing through the plane of this edge-on view are the dark areas of the Milky Way, such as the Great Rift, the dark nebulae, and the empty spaces between the spiral arms.

A bird's-eye view of the Milky Way.

An edge-on view of the Milky Way. In both views the cross marks the approximate location of our Sun.

THE MILKY WAY

As these spiral arms so strongly suggest, the entire Milky Way is in rapid rotation about its bulging center. It is this whirling motion that gives our galaxy its flattened disk-like shape. It is believed that our solar system is rotating around the center hub at a speed of about 150 miles per second, which amounts to one complete revolution in about 250 million years.

In looking at the Milky Way through binoculars on a sparkling, moonless night, you'll have the feeling that certainly those stars must be crowded close together, almost like a wheeling flight of blackbirds at roosting time in autumn. Knowing as we do that all these stars are in rapid motion, it seems inevitable that there would be collisions, for in our binoculars many of them appear to be almost in contact. This, of course, is all a matter of perspective. Each of the 200 billion stars in the Milky Way has its own home territory in a vast space 100,000 light-years across. This would seem to give each one ample room to roam around, for no one has ever witnessed a collision of two stars. Our own Sun, a typical star of the Milky Way, is not exactly crowded since its nearest neighbor star, Proxima Centauri in the southern skies, is some 26 million million miles away!

Of the incredible hordes of stars that form our galaxy, the number that we can see with the naked eye seems pitifully small. On clear, sparkling nights, I have sometimes asked visitors how many stars were visible to them. Their guesses have ranged from "thousands" to "millions" to "countless." Actually, the number that can be seen at any one time is about 1,500. On such occasions, I always remember the over-generous star counts of the little child in Robert Louis Stevenson's charming rhyme, "Escape at Bedtime":

> The lights from the parlour and kitchen shone out
> Through the blinds and the windows and bars;
> And high overhead and all moving about,
> There were thousands of millions of stars.
>
> There ne'er were such thousands of leaves on a tree,
> Nor of people in church or the Park,
> As the crowds of the stars that looked down upon me,
> And that glittered and winked in the dark.
>
> The Dog, and the Plough, and the Hunter and all,

And the star of the sailor, and Mars,
These shone in the sky, and the pail by the wall
Would be half full of water and stars.

They saw me at last, and they chased me with cries,
And they soon had me packed into bed;
But the glory kept shining and bright in my eyes,
And the stars going round in my head.

    The starry skies are also an escape for some who are no longer little children. After an evening of exploring the glories of the Milky Way through binoculars, even a seasoned watcher sometimes may put away the glasses and retire with the stars circling in his head.

CHAPTER 10
Variable Stars

When visiting a museum of natural history, I am sure that many of you have been particularly attracted to some cleverly wrought habitat exhibit showing groups of mounted animals just as they appear in nature. No doubt you found it quite an educational experience and a most convenient way to study them, for it could all be seen close up and in great detail. But it lacked the one thing that would have brought you back again and again to see it. It had no action to sustain your interest. It was a changeless, still-life scene.

Birdwatching is at its best when your feathered quarry can be observed busily engaged in some meaningful activity: feeding, nest building, courtship, or migration. Similarly, as a stargazer, the celestial specimens that are the most likely to capture and hold your attention and keep you on the edge of your observing chair are those stars whose light output is subject to change or fluctuation. These stellar show-offs are the variable stars.

From time to time in the earlier chapters devoted to the stars and constellations I called attention to certain of these inconstant stars that you could observe with your binoculars throughout either all or at least the brighter part of their cycle of change. The charts in this chapter show the naked-eye appearance of the sky field or constellation figure in which each of these selected stars is located. As with the seasonal sky maps, I have usually resorted to the use of connecting lines in order to make a star pattern or outline more readily recognizable. In nearly all cases, you can then find the variable itself (represented by an encircled dot (☉) symbol) near the center of a large circle that represents the immediate sky field as you will see it through 7x binoculars. This circle is ten degrees wide and includes stars down to magnitude 7.5.

Many of the charts also include a small inset view of the field around the variable. This is an enlargement made to extend the usefulness of the charts for those equipped with binoculars of higher power or small telescopes. On all charts, north is up.

On each chart you'll find the star's designation number, its period in days, and its magnitude range in brightness. The designation number is always a six-digit string that tells the star's location in the sky and helps you to plot its position on a star chart or atlas. For example, the designation number of the variable star Delta Cephei in the constellation Cepheus is 222558. The first four figures of this number place Delta Cephei at 22 hours, 25 minutes east of the vernal meridian of the sky. This is the star's right ascension, which corresponds to longitude in geography. The final two figures are the star's declination. They indicate that Delta Cephei is 58 degrees north of the celestial equator, an imaginary line passing around the sky directly above the Earth's equator. For stars south of the sky's equator, the two declination figures are underlined. Plot the right ascension and declination coordinates of an object on your atlas, then search in that region of the sky.

Nothing is more encouraging than a successful beginning. Just as in the early pages of this book I suggested that, as a good first step in stargazing, you go out and quickly locate the Big Dipper in the sky, now I suggest that you can just as easily become a variable-star observer if you take these charts, your binoculars, and a red-filtered flashlight out on the first clear night and find one or more of these variable stars for yourself.

No matter what the season of the year or the hour of the night, you are certain to find some of these fields in good position. You may already have taken this first step by using the naked-eye sky maps of the seasons. If so, do it again, for now you are almost certain to find that the star you saw before has changed in brightness. Such stars as Beta Persei and Delta Cephei are naked-eye variables that are always in your northern skies, so you can always find them with

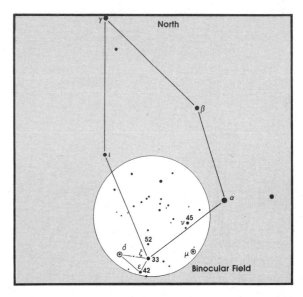

5⅓ Days   222558 Delta (δ) Cephei   Range 3.5-4.4

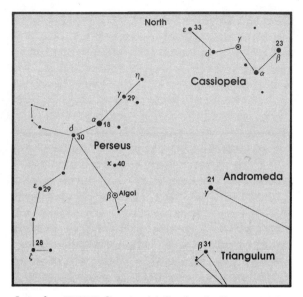

Irregular 005060 Gamma (γ) Cassiopeia Range: 1.9-2.8

binoculars — even on moonlit nights. With some of the others, however, don't be surprised if, after some searching, you locate the proper field only to find that the variable itself is missing. Don't be discouraged by its absence; it's simply proving that it's a variable star. Keep watching that empty spot and, sooner or later, you'll catch the first faint glimmer of the star's growing brightness.

I have drawn these charts with north at the top. This is correct for use with the naked-eye and binoculars, though not for most telescopes which invert images. Hold the chart so that its north point is toward Polaris, no matter whether the variable you are hunting for is in the east, south, west, north or even underneath the pole star. When properly held in this way, the chart and the sky field will be oriented the same way.

Once you have found the variable, it's a simple matter to estimate its brightness by comparison with stars that don't vary. On each chart you'll find an assortment of comparison stars with their magnitudes listed beside them. For example, look at the chart of the variable star 184205 R Scuti on page 89. Near the variable are three comparison stars marked 61, 67, and 71. These are magnitudes figured in tenths; I have omitted the decimal point so you won't mistake it for a fainter star. The star marked 61 is, therefore, magnitude 6.1, or just visible to the naked eye.

Now compare the brightness of the variable R with these three comparison stars. (Remember that with star magnitudes, the larger the number, the fainter the star.) In these three stars there is a spread of one full magnitude, from 6.1 to 7.1. Perhaps R looks fainter to you than 61 and brighter than 71. If it appears midway in brightness between the two, then its magnitude is 6.6. If not, then the star 67 will help you pinpoint your estimate.

Many variable stars, such as R Leporis and R Leonis, are decidedly red in color. Because these stars seem to grow brighter if you watch them steadily, estimating their magnitudes should be done by taking quick glances at them so you won't overestimate their magnitudes.

In general, variable stars have been named in the order of their discovery in the constellation in which they are located. A few, such as Algol and Mira, were given proper names long ago; others — for example, Delta Cephei and Eta Aquilae and Chi Cygni — have Greek letter symbols. But most have been given capital letter designations starting with the letter R. Thus R Leonis was the first variable to be found in the constellation Leo; Z would be the ninth. After Z was reached, paired letters were assigned beginning with RR as the tenth variable, then RS to RZ, SS to SZ, until ZZ was reached. After this came AA to AZ, then BB to BZ and so on until QZ was reached. The letter J was always omitted. This letter system accounts for 334 variables per constellation, but for the more prolific constellations, notably Sagittarius and Cygnus, even this wasn't enough. A number system was devised to follow QZ with V335, V336, etc. At latest count, Cygnus is up to V1815 and Sagittarius is up to V4091.

After you have located a variable star and made your estimate, the next important step is to keep a permanent record of its magnitude together with the date, hour, and any significant details of the observation. These estimates shouldn't be buried in a desk drawer where they can only serve as mementos of the past. They should be combined with those of other observers where they will be available to any interested researcher at any time. An estimate that you have made may just happen to be the only one in all the world made of a particular star at a given hour of a certain night.

Recently, I was able to furnish estimates of an old nova, 155526 T Coronae Borealis, which I had recorded fifty-five years before. And I still treasure my penciled record of the first estimate that I ever made. It was dated March 1, 1918, and was written on the ersatz note paper of that era.

If, in following the stars I have charted here, you begin to feel the first blissful birth pangs of an awakening interest in the systematic observing of variable stars, I urge you to write to:

> The American Association of Variable Star Observers
> 25 Birch Street
> Cambridge, MA 02138

The director of this worldwide association of observers will be pleased to send new members detailed instructions for making,

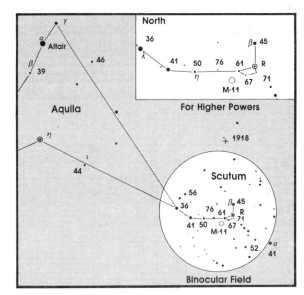

Semi-regular   184205 R Scuti   Range: 4.7-7.8

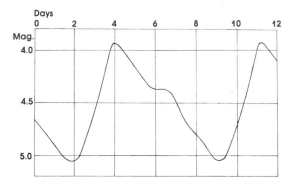

The Light Curve of Eta (η) Aquilae

VARIABLE STARS

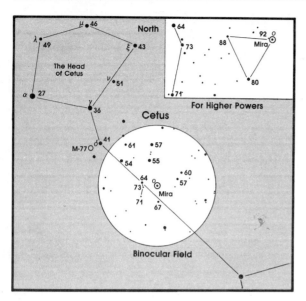

332 Days   0214<u>03</u> Omicron (o) Ceti   Range: 3.7-9.2

recording, and submitting observations. The list of newly found and still neglected variables is constantly growing, and there is still much pioneering work awaiting the earnest binocular observer, particularly in estimating and recording the changing brightness of semi-regular, eclipsing, and Cepheid variable stars. I suggest you make one particular member of this last class — a Cepheid variable — your first variable star. Like the Big Dipper, it is always in the sky and you can't fail to find it the first time you try.

Delta Cephei

This circumpolar variable star can be followed with your naked eye throughout its entire range. Binoculars are a great help, however, on moonlit nights or when the circling seasons or the passing hours have moved the "house of Cepheus" into that hazy region between Polaris and the northern skyline. Delta forms the apex of a small triangle of stars at the base of the "house." Its changing light can easily be noted by watching it on two or three successive nights. At maximum brightness it equals Zeta, its neighbor in the triangle; at minimum it is slightly fainter than Epsilon, its other neighbor.

Delta Cephei is the prototype of an entire class of variable stars known as Cepheids, named for the constellation Cepheus. Stars of this class are also called pulsating stars because of the strange manner in which their fluctuations are produced. It is believed that these stars alternately expand and contract much as our lungs do in the process of breathing, or the heart in its rhythmic beating.

Such stellar pulse rates are most precise, though by human standards they are quite prolonged: each heartbeat of Delta takes exactly 5 days, 8 hours, 47 minutes, 39 seconds. This is its fluctuation period, or the interval between one maximum and the next. It is the frequency of these heartbeats that has provided the solution to the age-old mystery of how far away many stars are. Knowing the fluctuation period and the apparent brightness of a Cepheid, an astronomer can determine the star's real luminosity. The difference between the actual luminosity and the observed brightness gives us the distance to the star.

Delta Cephei is some 275 light-years away, a fairly neighborly distance since it is a member of our own Milky Way galactic family.

Cepheids have also been found in remote star clusters and spiral galaxies, where their pulsing throbs have been recorded from as far away as half a billion light-years.

On the next clear night, make a house call and, with your binoculars as a celestial stethoscope, monitor the heartbeat of Delta Cephei.

Gamma Cassiopeia

This is a most eccentric star, one so unpredictable that no definite period can be assigned to it. In 1937, I watched it brighten to magnitude 1.6, nearly equal to Deneb in the Northern Cross (magnitude 1.3) and, therefore, definitely brighter than nearby Alpha in Perseus. After this surge, it slowly dropped to magnitude 3.0. It can usually be found near magnitude 2.5 but should be watched occasionally for any change. It has been suggested that Gamma is a rapidly rotating star that throws off fragments of its mass at intervals.

When estimating Gamma, use comparison stars that are at about the same altitude above the horizon as the variable so that your estimate isn't influenced by any difference in atmospheric density. Don't use nearby Alpha as a comparison star, for it has been suspected of being slightly variable.

The chart also helps you follow the changes in the eclipsing variable Algol in Perseus, already described among the autumn stars.

R Scuti

This is the star that I used earlier in explaining how to estimate magnitudes. It was chosen because it has such an accommodating array of nearby comparison stars. It lies in an easy field to find, for the bright cluster M-11 is in the same binocular field just to the left of the variable. Discovered in 1795, R was the tenth variable to be found in the entire heavens. It is a reddish star with a somewhat irregular period of 146 days. I have often found it at magnitude 5.0 but never fainter than 8.0. This is a brilliant region of the Milky Way that you'll find especially rich in your binoculars.

Eta Aquilae

Eta Equilae is a Cepheid variable very similar to Delta Cephei but with a slightly longer period and a greater range. It is located

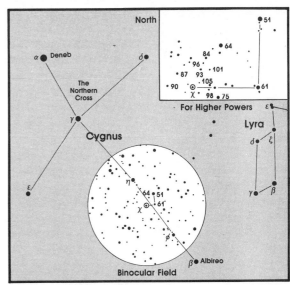

406 Days   194632 Chi (χ) Cyngi   Range: 5.1-13.3

436 Days   045514 R Leporis   Range: 6.0-9.7

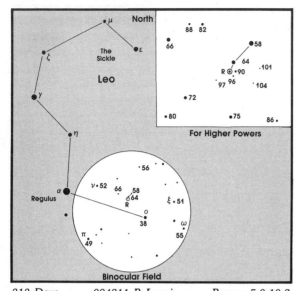

313 Days   094211 R Leonis   Range: 5.0-10.2

Irregular 154428 R Coronae Borealis Range: 5.9-15.0

just a twinkle north of the sky's equator. This star is characterized by a pronounced hump about midway in its descending light curve. If you plot your own light curve or graph of this star's changes, it will show a hump coming about two days after maximum when the fading light holds steady for a time. Eta and its comparison stars are shown on the chart of R Scuti.

Mira or Omicron Ceti

Mira is the "senior citizen" of the variable star clan — the first variable star discovered. It was first recorded by the Dutch amateur astronomer David Fabricius in August 1596 when it was at 3rd magnitude, or about equal to the faintest of the Big Dipper stars. A few weeks later it had disappeared. Johann Bayer, who gave the stars their Greek letter designations, saw the star in 1603 and, not knowing of its variable nature, assigned it the letter Omicron (o) in his atlas.

Since Mira has been under observation longer than any other variable, you might suppose that its behavior could be precisely forecast, but this is not the case. Both the date of maximum and the magnitude at maximum are subject to irregularities. I have never seen it quite as bright as its neighbor Beta Ceti, which is magnitude 2.0. As I write this, only two days before its predicted date of maximum, I estimate it at magnitude 3.2, which just shows that Mira doesn't go by the book.

Mira is a fine binocular variable. It can usually be followed through most of its range, though when it's faint, a nearby 9.2 star can be confused with it. Like Delta Cephei, Mira also heads a class of variables. These are known as Mira or Long Period stars because their periods range from 200 to 400 days. Mira is a red giant, somewhat like Betelgeuse in Orion, and it is immense — about 300 times the size of our Sun.

Chi Cygni

For a few weeks around the time of its maximum brightness, this variable is often conspicuous in the long arm of the Northern Cross where it may be found nearly on a line between Albireo and Gamma. It is classified as a Long Period variable since its period of 406 days is one of the longest of all. Discovered in 1686, Chi is one of the "old timers." Its range of ten magnitudes is also one of the most extreme, making this star some 10,000 times brighter at maximum than at minimum, though at some maxima the star doesn't reach magnitude 6.0.

R Leporis

This remarkable "Crimson Star" was first noticed in 1845. Its deep red color is quite pronounced in binoculars. It is another Long Period or Mira-type variable and, as you will note, its period of 436 days is even longer than that of Chi Cygni. You should be able to follow it through most of its cycle with 7x binoculars, though its low declination in the south may make its minimum phase rather difficult because of horizon haze.

R Leonis

This Long Period variable can be followed through all of its brighter range with binoculars; its reddish color sets it apart from the other stars in the field. R has always been a favorite star of mine, for it was the first variable I ever found. It's also one of the easiest to find.

R Coronae Borealis

On the basis of performance, this is one of the most interesting stars in the sky. It seems to be totally irregular and unpredictable with no definite period. Although it usually can be found at its maximum brightness of about magnitude 6.0 and has been known to remain there for ten years at a stretch, it can also suddenly drop out of binocular vision down to magnitude 14.5 and hover there for several months before gradually rising to 6th magnitude again. The spectroscope reveals that R has an excess of carbon in its atmosphere, and it has been suggested that this abundance may be a factor in the star's strange behavior.

R is readily found near the center of the circlet of the Northern Crown and is usually well within the power of your binoculars. Check on this star every night and keep a record of any changes that you see.

RY Sagittarii

This irregular variable has all the characteristic family traits shown by R Coronae Borealis, though it doesn't seem to fade quite

Irregular   191033 RY Sagittarii   Range: 6.1-14.0

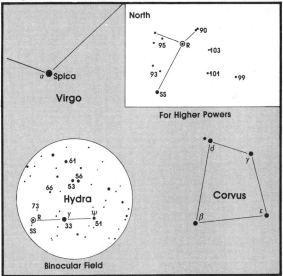

405 Days   132422 R Hydrae   Range: 4.2-9.5

VARIABLE STARS

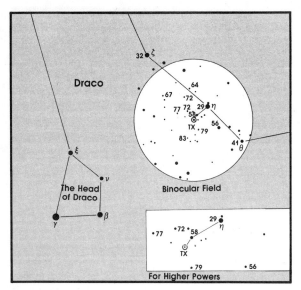

79 Days   163360 TX Draconis   Range: 6.8-8.1

so low at minimum. It also shows a slight secondary variation of about half a magnitude over a period of around forty days, though it takes a keen eye and good seeing to catch this. RY is just as interesting to watch as R Coronae Borealis although it is followed by relatively few observers. In our latitude, unfortunately, the star's brief sweep across the southern sky never carries it much above the haze of the horizon.

R Hydrae

The two lower bright stars in Corvus, the Crow, point directly to Gamma Hydrae, and R follows it by about two degrees. This is a Mira-type variable whose period has decreased by 100 days since its discovery in 1704. At maximum, this red star can be picked out by its color alone. Virtually all of its cycle is within range of your binoculars.

TX Draconis

TX Draconis is a reddish, semi-regular variable that is easily found, and its range of two magnitudes is always within your powers. In addition, it is always above your horizon.

CHAPTER 11
Novae and Nova Hunting

Novae are closely related to variable stars. The terms New Stars and Temporary Stars, which are often applied to them, are correct only with regard to their ephemeral visibility. These amazing stars, after their initial explosive outbursts, gradually diminish in light until, after a year or several years, they settle back to their original status as faint telescopic stars some nine to thirteen magnitudes below their maximum brightness.

The first recorded nova blazed out in Scorpius in 134 B.C. Its sudden appearance convinced the Greek astronomer Hipparchus that actual changes could occur among the stars. Six years later, he had completed a catalog of 1,025 of the brighter stars so that future observers would have a means of recognizing any variations that might take place in the skies to come.

In November 1572, the brightest nova on record appeared in the constellation Cassiopeia. It is usually known as Tycho's Star after its famous observer and recorder Tycho Brahe, who thus describes his first view of the spectacle: "One evening when I was considering, as usual, the celestial vault, the aspect of which is so familiar to me, I perceived with indescribable astonishment a bright star of extraordinary magnitude near the zenith in the constellation of Cassiopeia." According to Tycho, this star could be compared to the planet Venus when she is nearest the Earth. It remained visible to the naked eye for seventeen months.

The next bright nova in the skies was the so-called Kepler's Star of October 1604, which became brighter than the planet Jupiter and remained so for over a month. The summer key map (Chapter 7) shows the location of this long-ago explosion in southern Ophiuchus.

Following the appearance of Kepler's Star there were no novae of first-magnitude brightness for almost 300 years. This long drought came to an end in the opening year of this century with the discovery, on February 21, 1901, of a nova not far from Algol in the constellation Perseus. In less than four days, Nova Persei rose thirteen magnitudes to equal the brightness of neighboring Capella. Today it is again a 13th-magnitude star that occasionally shows fluctuations of a magnitude or more.

The evening of June 8, 1918, marked the debut of the brightest nova since Kepler's Star of 1604. That occasion has become one of the treasured memories of my own youthful nights as I was one of a host of other stargazers located all around the world who witnessed the performance. And I'm quite sure that at least one member of that far-scattered audience watched Nova Aquilae that night with the same "indescribable astonishment" with which old Tycho perceived his star more than three centuries before. This nova reached its maximum on June 9 when it equalled Sirius, the brightest of the stars. Although it has now been back at its old 11th-magnitude, pre-nova state for more than sixty years, for sentimental reasons I still take a peek at it from time to time.

In later years, other novae have paid brief visits to our skies. These always-welcomed guest stars came in 1920, 1925, 1934, 1936, 1942, and 1975, when a nova of magnitude 1.8 suddenly blazed out about 5 degrees northeast of Deneb in the Northern Cross. There, when darkness fell on the night of August 31, it was quickly spotted by scores of stargazers who were favored with clear skies.

Recurrent Novae

All of the novae thus far described have been content to make their one, brief burst of glory and then retire to pass their golden years in quiet obscurity. There is, however, a small group of these star performers who have made an encore and, in some cases, several repetitions of their act. U Scorpii has thus far exploded in 1863, 1906, 1936, and in 1979. RS Ophiuchi has equalled this with outbursts in 1898, 1933, 1958, and 1967. T Corona Borealis, normally a 10th-magnitude star lying just outside the horseshoe figure of the Northern

Crown, rose to 2nd magnitude in 1866; eighty years later, in 1946, it heard the call again and flared up to 3rd magnitude. These three recurrent novae are beyond the reach of your binoculars, except when they are bright, but I have marked the locations of the latter two on the key maps so that you may watch for any future flare-ups.

Supernovae

In the chapter on the autumn stars, I noted that a nova had appeared in the galaxy M-31 in 1885. That was no ordinary nova; it was a supernova. By definition, a supernova is one that is some 10,000 times brighter than the average nova. Every year a number of supernovae are found by photography in distant galaxies, particulary those in the Virgo and Coma Berenices fields. In some of these faint galaxies, the newcomer, when at maximum brightness, will give out more light than the entire galaxy in which it appears. Within historic times, three supernovae have exploded right here in our Milky Way galaxy. The first of these is said to have blazed out in Taurus in 1054 at the spot where we now find M-1, the Crab Nebula, a faint binocular object among the winter stars. Tycho's Star of 1572 and Kepler's Star of 1604 were also supernovae of our galactic system.

Nova Hunting

Nearly every year a nova bright enough to be seen with the naked eye is found somewhere in the sky, and many times it is the stargazing enthusiast who finds it. Here is a still uncrowded field for the persistent observer who sweeps the skies at every opportunity with both the naked eye and binoculars.

To the novice, the possibility of recognizing one more star among the host of permanent residents of the sky may appear to be quite hopeless, but in actual practice the procedure is greatly simplified. For one thing, the nova hunter doesn't have to search the entire sky but only that portion where game is most likely to be found. Just as the birdwatcher in search of waterfowl or shorebirds turns his steps toward lake or stream, so the seeker of strange stars directs his eyes or his binoculars along the mainstream of the skies — the Milky Way.

When we view the broad band of the Milky Way we are seeing

an edge-on section of our own galactic system, and here, almost in a single plane, we can scan most of its stars. The vast majority of recorded novae have appeared in a zone about 10 degrees wide lying on either side of the galactic circle, the mid-line of the galaxy. Our search for novae, therefore, instead of covering the entire sky, narrows down to a systematic nightly inspection of the slender strip of Milky Way that is above your horizon.

It goes without saying that to find a new star you should know the old ones. This can only be done by frequent exposure to them. In the pages devoted to the seasonal appearance of the skies, I have stressed the importance of learning the constellations not as formless scattered groupings of stars but as definite star patterns that you can quickly recognize each time you see them in the sky. Such patterns may well be of your own devising, with the principal stars of each group connected with imaginary lines to form some figure or design of your own choice.

Frequent inspection of these patterns will soon fix them in your mind. I find that as I sweep my eye along the Milky Way each of my star patterns passes in succession before me, each one so clear and definite from long association that any stranger within that design disturbs my mental image and automatically stops my eye. The culprit, be it nova, planet, or slowly creeping man-made satellite, is quickly located. Simply stated, if your star pattern is a square, a pentagon will tell you something is amiss.

Thus far, virtually all bright novae have been first sighted with the naked eye. Today, a systematic search for fainter novae is being made with binoculars in the hands of dedicated stargazers. In this work, G.E.D. Alcock, a schoolmaster of Peterborough, England, has been particularly successful. He began a program of sweeping for novae with binoculars in 1953 and now holds the record with four such discoveries in 1967, 1968, 1970, and 1976 — a fine example of patience and persistence.

Nova hunting with binoculars is naturally more difficult and time-consuming than with the naked eye because of the much smaller field of view and the greater number of stars that binoculars reveal. Here the naked-eye constellation patterns are of little or no assistance; the hunter must concentrate his efforts on a careful search of a small, specific area of sky. It may be a strip no wider than his field of view and no longer than his photographic memory. In this bounded bit of Milky Way he must again create his own star patterns to assist him in the prompt detection of a possible intruder.

Just what area or areas of the Milky Way the observer selects for his nova search program is entirely a matter of choice. Judging by past appearances, the region along the median line of the galaxy from Cygnus through Aquila, Scutum, and Sagittarius would be the most promising territory in which to stake one's claim. Nova hunting with binoculars is still in its pioneering stage in this country, and the prospector will find little competition for likely claim sites.

CHAPTER 12
The Sun

I can't write a book that claims to be a guide to the stars without paying tribute to the Sun — the nearest, brightest, and most essential star of all and the only member of that uncounted throng that we know, with first-hand certainty, to be the sole support of a large and varied family. Among these dependents are the nine major planets (Mercury, Venus, Earth, Mars, Jupiter, Saturn, Uranus, Neptune, and Pluto) as well as a multitude of tiny bodies known as asteroids. As second-generation members of the solar family, all but two of the major planets — Mercury and Venus — have families of their own in the form of moons or satellites that circle about them even as they revolve around the Sun. Finally, as part of this entourage are those far-ranging distant relatives, the comets and the meteors.

This family of the Sun is not noted for togetherness, for its members are scattered all about to a distance of well over four billion miles from the central hearthfire. The only tie that binds them is that of gravitation. Happily, nearly every member, save for the wayward planet Pluto and the smaller asteroids, can be captured with your binoculars. I shall begin a roll call of this family with the Sun.

That our Sun is only an average star is one of the facts of life that we somewhat reluctantly accept. This daytime star seems so vastly more important than the stars of night that we unconsciously regard it as being in command. I must confess that even after sixty years of stargazing I still don't class the Sun and stars together. Like Orion and the Scorpion, they never can appear together in the sky and thus they seem to be engaged in a never-ending struggle for supremacy. In the evening as the Sun sets in the west and the stars come out, it seems that they have vanquished it. Then in the morning we see the stars, in turn, fade in retreat, for now it is "the Sun which scatters into flight, the stars before him from the field of Night."

The Sun is a gaseous globe 864,000 miles in diameter, which is large enough to contain 1,300,000 Earths. If our Earth were placed at the center of the Sun, the Moon could revolve about it at its customary distance and still reach only a little more than halfway to the Sun's surface. Although the Sun is a mere pygmy compared to some of the red giant stars, such as Betelgeuse, Antares, and Mira, it is much hotter than these low-density stars. The temperature on its surface is about 11,000 degrees Fahrenheit, which is cool compared to the temperature at the Sun's center, an estimated 30,000,000 degrees.

For the binocular observer the most interesting feature of the Sun is the parade of sunspots across the solar disk as the Sun slowly rotates on its axis. Sunspots are believed to be magnetic storms on a scale so vast that many of them are far larger than Earth. These spots appear intensely black to the eye, but this is only by contrast with the brighter surface surrounding them.

Let me strongly emphasize that you should never observe the Sun directly through any optical instrument. The danger of permanent damage to your sight increases with the power of your glasses. The safest way to observe the Sun is to mount your binoculars on a sturdy tripod, remembering to cover one of the objective lenses so no one can accidentally look through it. Next, set up a piece of white cardboard several inches behind the other eyepiece so you can project the Sun's image onto the board. Once set up, you'll be able to see sunspots easily.

You will usually find some sunspots on your first look, though not always, for sunspot activity goes through a more or less regular cycle of about eleven years. At minimum, there may be days or even weeks when no spots, large or small, are visible. But at maximum, you may even find a double row of spots, both singly and in groups, scattered all across the solar disk. Spots are partial to the Sun's middle latitudes. They are never found near the poles and only rarely do

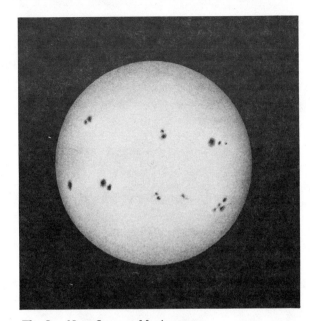

The Sun Near Sunspot Maximum

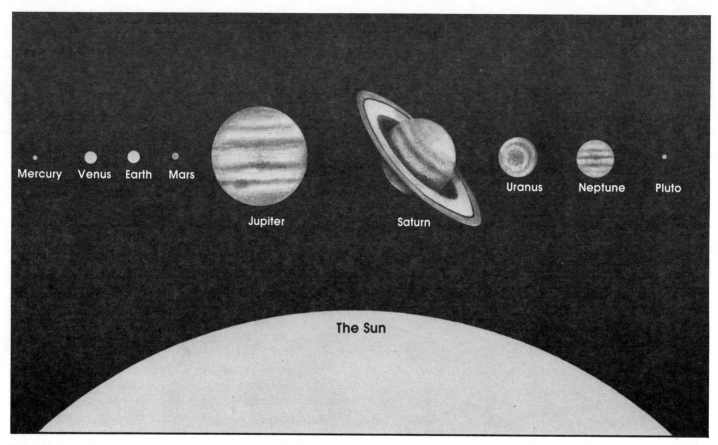
Relative Sizes of Planets and the Sun

they appear on the equator. Their latitude is subject to the sunspot cycle. At the beginning of a new cycle, the spots appear about 35 degrees north and south of the Sun's equator and gradually appear at a lower latitude as the cycle progresses.

Because the solar axis inclines 7 degrees to the plane of the Earth's rotation, the Sun's equator stretches in a straight line across the solar disk only in the months of June and December, when the Earth crosses the Sun's equatorial plane. As we view the sunspots from Earth, they seem to move across the Sun from east to west — always parallel to the Sun's equator. This is an apparent motion, however, for it is the Sun's slow turning on its axis that we really are watching. It was by timing these apparent sunspot movements that it was learned that the Sun could not be a solid body, for spots in different latitudes take different times to cross the Sun's disk. A spot at the Sun's equator makes a complete revolution in 25 days while one in mid-latitudes requires some two days longer.

Try sketching a large sunspot group as it appears through your binoculars on the cardboard. Follow it from day to day and carefully note its changes in form and size. Quite often such a group can be tracked all the way across the solar disk until it disappears at the western edge. If after a lapse of about two weeks it has survived its unseen passage across the Sun's opposite side, it will reappear at the Sun's eastern limb to begin another crossing. One exceptional group was observed for a year and a half, or some 36 crossings, though the average life of the larger groups is about two or three months. Many spots, however, will fade out and disappear even as you observe their initial crossing.

Should you wish to observe and record the life history of a sunspot group as it unfolds before you from day to day, be on the lookout for some of the following characteristics and peculiarities. A sunspot consists of a very dark nucleus called the umbra which, in a large spot, may be some 40,000 miles in diameter. This is bordered by a lighter zone known as the penumbra, which can be as much as 100,000 miles across. Often associated with sunspots are the faculae, which are exceptionally bright areas best seen when near the slightly darker edge of the Sun's disk.

A typical sunspot group begins in two parts about 50,000 miles apart with their centers lying in a nearly east-west line. Of the two parts, the western (on the right) is usually nearest the Sun's equator. Such a group may gradually shrink and disappear or it may grow still larger, in which case the eastern part may divide into smaller spots and slowly disappear while the western part becomes a large round spot that may persist for many crossings of the Sun.

In 1843, Heinrich Schwabe, an amateur stargazer of Dessau, Germany, after twenty-seven years of watching and recording sunspots, announced that the spottedness of the Sun follows a periodic cycle that has been determined to be about eleven years. It is near the times of maximum activity — when a large sunspot group crosses the Sun's meridian and is turned directly toward us — that the Earth is most likely to have magnetic storms and active displays of aurorae, "northern lights," as well as unusually strong radio interference. The last such sunspot maximum was around 1980.

If in exploring the night skies you may have wondered what some star, such as Capella, might look like if seen close up, just turn your darkened glasses on the Sun. What you see there is a small Capella that is so near at hand that you can feel its heat; its disk is so bright that it completely drowns out every other star.

I heartily recommend sunspot watching as a fascinating and challenging trail for the serious observer to follow with binoculars. In the broad sense, this is stargazing just as truly as sweeping through Orion or the Milky Way. It is Stargazing spelled with a capital S for Sun.

The American Association of Variable Star Observers (address given in Chapter 10) sponsors a Solar Division for those interested in observing and recording sunspots.

CHAPTER 13
The Planets

The previous chapter on the Sun listed the names of the major planets in the order of their distance from the Sun. With the aid of binoculars, we will get a closer look at each of them with the exception of small Pluto, which is just too little and too far away. An acquaintance with our brother and sister planets is more than just a matter of family pride and kinship, it is a worthwhile accomplishment that a stargazer will be able to use to recognize the bright wanderers that so frequently disturb the familiar constellation figures.

It is, I admit, a bit more difficult to get to know the planets than their "look alikes" among the stars. This is because the stars are fixed in place; they never leave the star pattern that surrounds them. Antares, for example, will always be the heart of the Scorpion, whose long, curved tail dangles down to the horizon, but the planet Mars — often Antares' identical twin in color and brightness — may appear in any of a dozen constellations. Aldebaran never leaves its familiar place in the V-shaped Hyades, but the planet Saturn, which equals it in brightness when its rings are turned edgewise toward us, has no permanent star pattern that will serve to point it out. The fact that a star twinkles while a planet shines with a fairly steady light may be of some assistance in identification, but even a planet will twinkle somewhat when near the horizon. The best guide of all is a thorough knowledge of the brighter stars that lie along the interconstellation route that is traversed by the planets.

This planetary pathway is called the zodiac, a narrow strip of starry background that extends all the way around the sky and

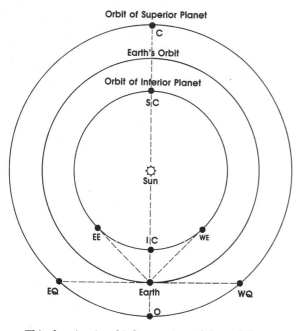

This drawing is a bird's-eye view of three different planetary orbits. The inner circle represents the orbit of an inferior planet, such as Mercury or Venus, that revolves inside the Earth's orbit. The outer circle is the orbit of a superior planet, one that revolves outside our orbit.

An inferior planet cannot be seen from Earth at either IC or SC. We view these planets best between IC and EE (after sunset in the western sky) or between IC and WE (before sunrise in the eastern sky).

A superior planet cannot be seen at C. Such a planet is nearest the Earth and therefore brightest at O when it is opposite the Sun in the sky, though it can be well observed from EQ to WQ when it is at right angles to the Sun.

against which the Sun, Moon, and planets appear to move as we view them from Earth. This band of sky is 18 degrees wide with the ecliptic, the apparent path of the Sun, as its centerline. Only the Sun and the Earth, as viewed from each other, stay strictly on this centerline, for the ecliptic is the plane of the great circle which the Earth draws around the Sun each year. Each of the principal planets, as well as our own Moon, travels in a plane tilted more or less to the ecliptic but always within the borders of the zodiac.

Mercury

In any planetary book of records, the planet Mercury would be a frequent entry. It is the nearest planet to the Sun — some 36 million miles distant. Along with Venus it is one of the hottest planets, with a temperature of some 800 degrees Fahrenheit on the sunward side, but the nighttime side may well be the coldest of all at −280 degrees. With a diameter of 3,000 miles, it's the smallest of the major planets, and its orbital speed about the Sun of 30 miles per second makes it easily the swiftest. Radar observations indicate that Mercury's day equals 59 of our days and, until Pluto came along and challenged it, Mercury also claimed the most eccentric orbit and the orbital plane most inclined to the ecliptic.

Mercury is the most elusive of the brighter planets; many who are otherwise quite familiar with the sky have never seen it. This is because in our temperate latitudes it can never be seen against a completely dark sky. Even at the times of its greatest elongations east or west, it is never more than 28 degrees distant from the Sun, which is less than three wide-fields in binoculars. Nevertheless, in spite of these restrictions finding Mercury can be a fairly simple procedure when the time and conditions are the most favorable. I have often seen it with the naked eye glowing like a distant beacon in a clear western twilight sky less than an hour after sunset in early spring. When well seen, Mercury can be conspicuous, for at its brightest it appears to the naked eye almost as bright as Sirius, which at times is also shining not far distant. Most often, however, Mercury can be found at about the brightness of Capella or Arcturus.

Apparitions of Mercury are particularly good when they occur in the western twilight sky in the months of March and April and in the morning sky before sunrise in September and October. Note the point on your horizon where the Sun sets or rises; this will be one point on the ecliptic. At the time of an elongation, you should be able to easily find Mercury with binoculars not far east or west of this point.

In a number of ways Mercury is quite similar to our Moon. Though it is somewhat greater in size and density than the Moon, it, too, is a poor reflector of the sunlight that falls on it. Space probes have shown that its surface is a pitted, crater-pocked terrain much like the lunar landscape. Mercury also has phases which are, in miniature, just like those of the Moon, though they can't be observed with low-power glasses. Finally, Mercury sometimes comes directly between Earth and the bright face of the Sun, as does the round ball of the Moon.

These infrequent transits across the Sun provide the greatest thrill for the binocular observer of the planet. To prevent eye injury, the same precautions must be taken here as in observing sunspots. When in transit, the little planet looks much like a perfectly round sunspot without the usual penumbra. Like a sunspot it moves across the Sun from east to west since transits occur only when the planet is in retrograde, or westerly, motion. About thirteen transits of Mercury occur each century, and these can only take place in May and November. The last transit was on November 10, 1973, and the next is November 13, 1986. There will be two more in the present century on November 6, 1993, and November 15, 1999. Each transit takes about five hours. I wish you cloudless skies for these occasions.

Venus

Venus, the second planet from the Sun, is very nearly Earth's twin sister, for she is only slightly smaller in size, mass, and density. She is also our nearest planetary neighbor, at a distance of 26 million miles. Like Mercury, Venus shows phases as she travels around the Sun, and in this case they *can* be seen in binoculars. The crescent phase is most easily seen when the planet is about 35 days before or after inferior conjunction, when it passes between the Earth and Sun. If you follow the course of the planet with your glasses, you'll notice a great variation in the size of the disk as it travels around the Sun. This is because Venus is six times farther away from us at superior conjunction than she is at inferior conjunction, when she comes closest to Earth.

At her brightest Venus is at magnitude -4.5; next to the Sun and Moon it's the brightest object in the sky — a fact that for years has made her a prime target for those who have a penchant for seeing flying saucers. Because of her great brightness, you will probably want to observe Venus in daylight or twilight rather than at night. It isn't difficult to locate this planet in afternoon daylight if you mark its course the day before during evening twilight with fixed terrestrial objects such as trees and buildings and then extend this course 15 degrees eastward for every hour before twilight begins. A still simpler method is to get the planet in your binoculars in the eastern sky before dawn when the sky is still dark. With your glasses mounted on a tripod you can follow the planet long after the Sun is high up in the sky.

Despite its closeness to us, no one has seen the surface of Venus from Earth — even with the most powerful telescopes — for it is always covered with a heavy layer of clouds. Since no surface markings are visible, its rotation on an axis cannot be timed. But soundings with radar indicate that its day is 243 of our days in length, which is longer than a Venusian year of 225 days required for its circuit of the Sun. Also of interest, Venus is the only planet that rotates on its axis in a clockwise direction.

Space probes have now shown that the surface temperature of Venus is about 900 degrees and its atmosphere is made up mostly of carbon dioxide. It is definitely no place for stargazers, as the thick cloud cover would prevent them from seeing any stars or even their near neighbor, Earth.

Venus, like Mercury, occasionally transits the Sun, and at such times its fairly large black spot can be easily seen with the naked eye. You'll have plenty of time to get your Sun filter ready for the next transit since it won't occur until June 7, 2004. It is doubtful if any astronomer living today has ever witnessed a transit of Venus as the last one took place in 1882. In connection with the extreme rarity of these events, the following note concerning the activities of a couple of young stargazers of long ago may be of interest:

> The first authentically observed transit was one of Venus, which occurred in 1639. It was predicted by Jeremiah Horrocks, a young English clergyman, and as it occurred on a Sunday, his ministerial duties prevented him from seeing the beginning of the transit; but his friend Crabtree, whom he had informed of the impending phenomenon, witnessed the entrance of the planet on the solar disc, and both young amateurs saw it before the end. It was seen by no other observers.

I like to picture in my mind the tense drama taking place in the little Lancashire village of Toole on that Sunday afternoon as young Jeremiah, then but twenty years old and serving his first pastorate, hurried through the discharge of some duty in connection with his ministry, while all the time his thoughts kept straying to the clearness of the skies. From time to time he would look up anxiously as drifting clouds dimmed the sunlight streaming through the stained glass windows of the church. Those same clouds could prevent his witnessing a sight that no eyes in all the world had ever yet seen. His mind was also on the flying minutes and this added to his worries for, according to his calculations, at three o'clock that afternoon the small black image of the planet would start to creep across the Sun. Right now, he told himself, this transit may already be in progress.

When he finally arrived and got the Sun in focus, Jeremiah found the planet already on its westward trek across the bright face of the Sun. He was able to watch the transit for half an hour until the Sun set on that brief but momentous December day. Needless to say, young Horrocks was overjoyed that his calculations had proved to be completely accurate and that he had witnessed that rarity of rarities — a transit of Venus. He knew full well that he would never see another, for the next would not occur for more than 120 years.

On that Sunday afternoon in 1639, two centuries before the days of Darwin and *The Origin of Species*, we have an early example of the conflict between religion and science. And it was solved quite happily — by compromise.

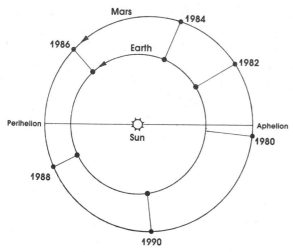

Oppositions of Mars 1980-1990
The straight lines drawn between Earth and Mars show the relative distance between the two planets at time of opposition. The precise dates for these oppositions are: Feb. 25, 1980; Mar. 31, 1982; May 11, 1984; July 10, 1986; Sept. 28, 1988; Nov. 27, 1990.

Mars

Mars, the fourth planet from the Sun, has a diameter of some 4,200 miles, a little more than half that of Earth. In spite of its small size, it becomes, under favorable circumstances, the brightest object in our night skies after the Moon and Venus. In binoculars, the red color of the planet is quite pronounced, though with low power it is difficult to make out any definite disk.

Mars revolves about the Sun in a path that lies outside the orbit of the Earth and, therefore, we can never see it in a crescent phase as we do Mercury and Venus. Should you ever find Mars comparable to Jupiter in brightness, it means that our neighbor planet is near opposition, the point where the Earth is almost between Mars and the Sun. It is at opposition that Mars comes closest to the Earth in its 687 day trip around the Sun. However, there are oppositions and then there are *oppositions*. Because the Sun is not at the center of Mars's orbit, the red planet comes nearer the Earth at a perihelic (nearest the Sun) opposition than at an aphelic (farthest from the Sun) opposition.

The distance between Earth and Mars can vary from about 36 million miles at the very best oppositions to about 63 million miles at the least favorable. On the night of August 11, 1971, Mars was at its nearest and was brighter than the planet Jupiter. Such a close approach won't occur again until the year 2003, though fairly good oppositions will come in 1988 and 1990, as my drawing shows.

However, if you observe from a northern vantage point these close oppositions of Mars may not be as advantageous as some of the more distant ones because the planet is always well south of the equator. At the close opposition of August 1971, Mars was at 20 degrees south declination so that we in northern latitudes viewed it through a much thicker blanket of Earth's atmosphere than in 1973 when it was 10 degrees north of the equator. Astronomers can overcome this perversity of nature by building their observatories in such favored locations as Peru, South Africa, or Australia, where the action takes place almost directly overhead.

At any opposition it is interesting to follow closely Mars' nightly trip against its starry background. As the time of opposition approaches you'll notice that the normal eastward movement of Mars among the stars is slowing down. Eventually it will seem to come to a complete stop, then reverse its direction and move backwards, or toward the west. After a few weeks this backward, or retrograde, motion slows to a halt and Mars turns eastward, resuming its normal motion. The explanation of these strange didoes in the sky is a simple one. They occur at a time when Earth and Mars are nearest. Earth, being nearer the Sun, moves more rapidly in its orbit than Mars and, at opposition, is passing Mars, thus making the red planet seem to be moving backward against its starry background — just as a speeding car, in passing a slower one on the highway, makes the latter seem to be moving in reverse against the landscape in the distance. If you plot the motion of Mars during its retrograde phase you will find that, due to the inclination of its orbit, it makes a slender, well-defined loop as seen against the stars.

This demonstrates the great advantage of binocular over naked-eye observing. The glasses reveal a background liberally sprinkled with stars among which the planet clearly shows every detail of its changing path, its slowing motion, its acceleration, and the brief stationary pause at either end of the narrow retrograde loop.

Mars rotates on its axis in 24 hours 37 minutes 22.58 seconds. The split-second figure testifies that the planet has an atmosphere of sufficient clarity to permit some of its surface markings to be measured with great accuracy. It's this same transparency that has made Mars the most critically observed as well as the most controversial of all the planets. Over the past century, a number of keen-sighted observers, both in America and Europe, have claimed to see a network of dark lines geometrically placed all about the circumference of the planet. These narrow lines, both curved and straight, came to be called "canals" and were assumed by some to be artificial waterways constructed by a race of intelligent beings in an effort to conserve and distribute the water from the seasonal melting of the snow or ice in their polar regions.

Among those who claimed to clearly see these markings was a wealthy amateur astronomer, Percival Lowell of Boston. The brother of poet Amy Lowell and A. Lawrence Lowell, president of

Harvard University, he built and equipped a fine private observatory in the clear skies of Flagstaff, Arizona. From this mile-high station soon came drawings showing hundreds of vegetation-bordered canals and oases in great detail. Around the turn of the century, Lowell published three books that dramatically described Mars as the abode of an advanced civilization. Few astronomers of the time were in agreement with these theories and purported sightings. Professor E.E. Barnard, who was said to have had the most acute eyesight on record, was never able to see a single canal. But the public loved the Lowell books and the case for the canal-digging Martians got a terrific boost.

Inspired perhaps by Lowell's claims of a nearby neighbor culture in our solar system, another Mars-related book appeared in this same era that was to have an even greater impact on both the reading and listening public. In 1898, in the midst of the canal controversy, English novelist H.G. Wells published his early science-fiction thriller, *War of the Worlds*, in which an armada of space ships from Mars arrives on Earth and begins to take over our planet. We had no defense against the sophisticated weapons of the invaders, but neither had they any resistance to our germs, and so they perished to the last Martian.

War of the Worlds was Earth's first science-fiction involvement in interplanetary warfare, and it also proved to be the most realistic when, forty years later on the night of October 30, 1938, the story was dramatized on radio by Orson Welles. Fortunately, I heard the complete program from the beginning and knew that it was fictional in nature, but for many who tuned in late and missed the opening announcement, it was a night of sheer terror — especially for those living in New Jersey where the supposed invaders landed. It was a case of Welles interpreting Wells a bit too well.

As it finally turned out, about thirty years after this frightening episode, it was space ships *from Earth* that completely vanquished the Martians as well as their canals. Beginning in 1965, a series of three Mariner space probes made close flyby approaches to Mars, and in 1971 Mariner 9 was placed in orbit around the planet. The many photographs taken by these unmanned probes showed a surface not criss-crossed with interconnecting canals as Professor

Lowell would have had it, but thickly strewn with craters much like those that spot the surface of the Moon.

As a final blow to the Martian cause, the prying Mariners also found an extremely low temperature, no surface water, and an atmosphere composed of 95 percent carbon dioxide. Two later Viking vehicles soft-landed on the surface but found no organic matter in the soil.

In 1877 two tiny moons were discovered circling closely about Mars. Phobos, the inner one, is only 5,700 miles above the planet's surface and crosses its skies three times a day from west to east. Deimos, the outer moon, is some three times more distant from the planet. Both of these moonlets appear on Mariner photographs as heavily cratered masses of rock or metal that are roughly potato-shaped rather than spherical. Phobos is about 16 miles in length and Deimos is only half as large. Both are probably captured asteroids.

Mars and Its Two Moons
Phobos (left) and Deimos (right)

The Asteroids

If you were to make a scale model of the solar system with the nine planets properly spaced according to their distances from the Sun, you would find a large gap between the planets Mars and Jupiter. According to astronomical logic, there should be a planet somewhere in this void. Its apparent absence so worried a number of observers in the late 18th century that they made a concerted effort to find the missing body.

On January 1, 1801, Sicilian astronomer, G. Piazzi, while engaged in mapping the stars, found what proved to be at least a part of the unknown planet. It was not really much of a planet, being only 635 miles in diameter, but it was at the proper distance from the Sun and it pursued the proper path. It was given the name Ceres from the mythological goddess of agriculture, not from the fact that it proved to be the first of a *series* of four of these tiny planets, or asteroids, discovered between 1801 and 1807.

Ceres remains the largest of the asteroids. It was followed by Pallas (diameter 335 miles) discovered in 1802, Juno (diameter 155 miles) in 1804, and Vesta (diameter 340 miles) in 1807, after which no more were found until 1845. Today asteroids are still being found on photographic plates, and well over 2,000 are now known and cataloged. Most of them are quite small, ranging in size from small mountains to irregularly shaped rocks less than a mile in diameter. In many cases their orbits are quite eccentric. For example, on October 28, 1937, tiny Hermes, only a mile in diameter, came within 485,000 miles of the Earth; it could, on some future pass, come even closer to us than the Moon.

Of the brighter asteroids, Ceres, Pallas, and Vesta are easily seen with binoculars. I have seen Vesta, the brightest of them all, with my naked eye without difficulty. The astronomical periodicals often include charts and positional data showing their locations in the sky.

Jupiter

This giant of the planets is larger than all the other planets combined. Its mass is 300 times that of Earth, and its mighty globe could contain 1,300 bodies the size of Earth. It is the fifth major planet from the Sun and is more than three times farther from it than its red neighbor, Mars. Jupiter moves along its orbit at the leisurely speed of only 8 miles per second, which results in a year that is the equivalent of twelve of ours and conveniently places the planet in a different zodiacal constellation every year.

In sharp contrast to its long year, Jupiter has the shortest day of any of the planets, for it rotates on its axis in 9 hours 55 minutes. This is indeed surprising when you consider the immense size of the globe. Any spot on the equator of Jupiter travels at a speed of some 28,000 miles per hour. Through the ages this has generated sufficient centrifugal force to make the giant planet into a globe with a bulging equatorial belt and noticeably flattened poles.

Quite unlike Mars, we never see the actual surface of Jupiter; we see only the top layers of the clouds of hydrogen compounds that cross the planet in narrow belts parallel to the equator. By careful observations of these belts with powerful telescopes and space probes astronomers have learned that the equatorial belts make a complete rotation about five minutes faster than the belts nearer the polar regions of the planet.

For an observer with binoculars, Jupiter is probably the easiest of all the planets to distinguish, for he carries his identification marks in the form of four moons that circle about him in an ever changing retinue of shadows, transits, and eclipses. These four moons, discovered in 1610 by Galileo with his tiny, brand-new optic tube, are all within the powers of your steadily-held binoculars, and you can find delight and fascination in following their activities from night to night as they play slow-motion hide and seek about the bright globe of the planet.

In keeping with Jupiter's size, these four moons are also sizable affairs, although your binoculars will show them only as tiny moving stars. In the order of their distance from the planet, they are: Io, Europa, Ganymede, and Callisto. Io, though a little larger than our Moon, is the most difficult to see as it is the closest to the glare of the planet. Europa, the smallest of the four, is slightly smaller than our Moon. Ganymede, the brightest of them all, is somewhat larger than the planet Mercury, and Callisto, a rather poor light reflector, is a little smaller.

The orbits of these four bright satellites lie almost in the plane of Jupiter's equator which also is virtually in the plane of the ecliptic, so that we see the satellites nearly in a straight line drawn through the equator of the planet. While watching the transits, eclipses, and occultations of these four moons as they circle about Jupiter, you may realize that you are watching an animated miniature enactment of what you might be seeing if you stood on Jupiter watching the slow circling of the four inner planets — Mercury, Venus, Earth, and Mars — about the Sun.

It has occasionally been reported that Jupiter's moons have been seen with the naked eye. I find it easy to believe such claims, since their magnitudes range from 5.0 to 6.0. Such sightings would present no problem if the moons were located farther from the glare of Jupiter. It would make a most interesting project to see if you could spot any or all of these satellites with your naked eye. Make a drawing of their locations in relation to the planet, then check with your binoculars to learn if you really did see them. This is a difficult test of eyesight, so to give yourself every advantage choose a time when Jupiter is near opposition in one of the northern zodiacal constellations such as Taurus, Gemini, or Cancer and at a hour when the planet is near the meridian. It will then be at or near its highest point in your sky and can be seen through the least possible atmosphere.

The Jovian satellite system is truly an enormous one, for the planet actually has at least 16 or 17 known satellites, though only the four bright moons are within binocular range. The others are all small, not over 100 miles in diameter, and can be seen only with powerful instruments. In connection with these satellites, the following paragraph, which I found in a book on astronomy published in 1914, is of historical interest:

> Jupiter has 8 moons which have been discovered. The method of searching by photography is so nearly perfect now that it is extremely doubtful that more will be found.

The fourteenth moon was found, by photography, in December 1975, and several others have since been discovered by ground-based observatories and spacecraft. The Voyager craft even discovered a faint ring around the planet! That which was incredible yesterday may be commonplace today, and what was once thought perfect has been improved on since.

In addition to providing uncounted hours of fascinating entertainment for generations of sky observers, Jupiter and his four bright moons made a definite contribution to exact physical science when, in 1675, the Danish astronomer, Olaus Roemer, noted a discrepancy between the predicted and observed times that satellites were eclipsed in Jupiter's shadow. He found that these times varied with the distance between Earth and Jupiter: when the Earth was closest to Jupiter eclipses occurred 16 minutes earlier than eclipses that took place when we were on the opposite side of our orbit, some 186 million miles more distant. Those 16 minutes were the interval that light requires to cross the orbit of the Earth, making the speed of light a little more than 186,000 miles per second.

You will find the configurations and the nightly phenomena of these four bright satellites of Jupiter conveniently plotted for you in astronomical periodicals and in a number of observer's handbooks.

Saturn

Saturn is the second largest of the planets and the sixth in order of increasing distance from the Sun. When seen for the first time in a telescope of fair size and performance, it never fails to evoke an exclamation of amazement and delight from even the most blase, for, when well posed and properly positioned, it is easily the most pleasing picture in the solar system's gallery of skyscapes. Unfortunately, when seen in low-power binoculars, it is apt to be a disappointment.

I am sure that old Galileo knew a measure of this same chagrin when he turned his tiny telescope to this outpost planet on those nights in 1610, for he failed to figure out the true nature of the strange three-part shape he saw. When it later changed into a single ball, he gave up in despair and turned his talents elsewhere. In probing this same planet today, we shouldn't expect too much either from our eyes or our binoculars. The songbird in a nearby treetop is, quite understandably, an easier conquest than is a tiny gleam of light 900 million miles away.

Like Jupiter, Saturn has a rapid rotation on its axis with a day that is only about 45 minutes longer than that of the giant planet. As a result, it has the characteristic flattened poles and bulging equatorial girth. It also is well supplied with satellites. Along with the nine moons that have long been known, at least thirteen others have been discovered in the last decades — either from the ground or with Pioneer and Voyager spacecraft. Of these, your binoculars may give you just a glimpse of 8th-magnitude Titan, the largest, on a winter night of perfect clearness when Saturn is high up in the sky. Titan is somewhat larger than Mercury and may be the largest of all moons.

Saturn requires 29½ years to travel completely around the Sun at an orbital speed of six miles per second. This leisurely pace keeps it wandering through the same zodiacal constellation for three or four years at a stretch. The density of the planet is less than that of water, so it would easily float in an ocean large enough to contain it. Its polar axis is inclined to the plane of its orbit even more than our axis inclination of 23½ degrees; therefore, Saturn, too, has its seasons. And this, as you will see, is most fortunate for Saturn-gazers on Earth.

It is because of this tilt in its axis that we're treated to three-dimensional views of Saturn's most famous feature, its unique ring system. It consists of what appears to be three concentric, flattened rings fitted together, one within the other, completely encircling the planet some 9,000 miles directly above the equator of Saturn's 75,000-mile-diameter globe. The Voyager craft revealed that these rings are actually composed of hundreds of rings. This system of rings is 170,000 miles across but less than 10 miles thick. If the length of this printed page were to represent the diameter of Saturn's ring system, the thickness of this paper is four times too great to be in scale with the extreme thinness of the rings!

During half of Saturn's long trip around the Sun, or about fifteen years, the north side of the planet and the rings are tilted toward the Earth. As the season advances more and more of the ball comes into view. The rings appear to flatten out until, at midpoint, we see the planet from pole to pole and the ring, for a brief time, completely disappears. After this, the system will tilt away from us and the opposite or underside of the rings and the other pole will come into view for the remainder of Saturn's year.

My last ringless view of Saturn came in 1966. I watched it through a 12-inch refractor, and according to my records, the rings had completely disappeared when I looked on the night of October 29, though I still could see the black shadow that they cast across the equatorial belt of the planet. When clear skies came again on November 1, the rings had reappeared, though not yet as a solid line but as a series of bright dashes broken by dark gaps where the ring divisions lay.

When opened to their greatest extent as seen from Earth, the rings reflect more than twice the light as the ball of the planet; therefore, Saturn's brightness as we see it is governed largely by its ring position. At its brightest, the planet may equal Capella or Arcturus, but without the ring, when the edge is turned toward us, Saturn will be no brighter than Aldebaran or Altair. The rings are neither solid nor fluid but are composed of small particles such as rocks or other aggregate revolving independently about the planet.

Among my acquaintances are birdwatchers who have traveled

a thousand miles to see a Kirtland's Warbler in its native haunts. The night sky has its own rare specimens that are well worth a visit to an observatory, or perhaps an evening with a friend who has a telescope can be arranged. No one should pass through life without a look at Saturn and his rings.

Uranus

All of the planets described thus far have been known since ancient times, for they are easily visible to the naked eye, but far outside the orbit of Saturn slowly revolve three major planets — Uranus, Neptune, and Pluto — that came to us only through telescopes.

The first of these, Uranus, was discovered quite by accident by an amateur stargazer and telescope maker who was best known among his acquaintances as an excellent oboe player in a professional orchestra.

Like countless amateurs have done since, William Herschel built a telescope. With this six-inch instrument he began to poke around the skies and, perhaps unlike the great majority of his followers, he kept a careful record of his sightings. For example, in his journal for the night of March 13, 1781, his entry was a strange note, even for an oboe player: "In the quartile near Zeta Tauri, the lowest of two is a curious either nebulous star or perhaps a comet."

For nearly a year the object was observed as a comet, but finally it was shown to be traveling in a nearly circular orbit that could only be that of a planet. Thus Uranus, the god of Heaven, was welcomed into the official family of the Sun, and William Herschel was duly knighted by King George III as the discoverer of a New World.

But just as the "New World" of Columbus had been sighted many times before by others, so too had Uranus been seen and recorded at least a score of times by four different observers before Herschel came upon it. John Flamsteed, England's first Astronomer Royal, saw it on six occasions from 1690 to 1715, and Pierre Lemonnier, a French astronomer, observed it on twelve nights between 1750 and 1771. Each time it had been recorded as an ordinary star.

Uranus is the fourth largest planet and the seventh in order of distance from the Sun. Its axis is the most highly inclined of any planet, about 98 degrees; thus, when the polar axis points toward the Earth, it gives the planet the appearance of rolling along its orbit rather than spinning on its axis like a top.

Until recently, Uranus was known to have only five moons. Herschel had found the brightest two on a single night in 1787. But the count tripled over a span of three short weeks in January 1986 as the Voyager 2 space probe flew by the planet and discovered ten new moons, bringing the total to fifteen. These new moons were all tiny compared to the five previously known moons — Titania, Oberon, Umbriel, Ariel, and Miranda — the smallest of which is three times larger than the largest of the newly discovered satellites.

Voyager 2 has been exploring the solar system since 1977, sending back to Earth invaluable pictures and data from Saturn, Jupiter, and now Uranus. The photos of Uranus showed a uniformly colored sphere with only the slightest hints of features within the blue-green atmosphere. Voyager also photographed the nine, thin rings circling the planet and revealed a new tenth ring. The length of a Uranian day was also accurately determined from data returned by Voyager to be approximately 17 hours. Voyager's next, and last, planetary rendezvous is with Neptune, which it will reach in August 1989.

All along Uranus has been reluctant to reveal its secrets. Not until February 1948 was its fifth moon found, and as recently as March 10, 1977, it was discovered that Uranus, much like the planet Saturn, has rings. William Herschel would have readily accepted the announcement of these rings, for at one time he thought that he had seen such things around his planet. He never would have believed, however, the way in which they were discovered: through a 36-inch telescope in an airborne observatory eight miles above the Indian Ocean!

Like the planet itself, the rings were found by accident. Their discovery was a by-product of an effort made by NASA's Kuiper Airborne Observatory to record photoelectrically the occultation by Uranus of an 8.9-magnitude star, SAO 158687. Forty minutes before the star was due to be occulted by the body of the planet its light suddenly disappeared for seven seconds and then recovered. Several minutes later the star's light again blacked out and again recovered. At shorter intervals, this fluctuation was repeated until a total of five minima were recorded.

After the fifth dip in light came a period of 30 minutes in which the star's light continued steady until it was occulted by the planet. When the star reappeared 25 minutes later there followed another period of 30 minutes of steady light followed by another series of five disappearances, but this time their intervals were spaced in *exactly the reverse* order. It could only mean that the shadows cast by five concentric rings had passed in succession across the open telescope. Further observations have confirmed these rings and revealed four more.

When seen at its best, at opposition, Uranus is visible to the naked eye at magnitude 5.7 and, of course, it can be found without difficulty with binoculars if you know its approximate location in the sky.

Neptune

Scarcely had Uranus firmly established his true status as a new world when astronomers found him straying somewhat in his path around the Sun. He was seldom precisely where their computations showed he should be. For years he moved faster in this orbit than it seemed he should, then, as though weary, he began to lag behind until, only some sixty years after discovery, the planet was off its predicted position by an apparent distance equal to about one-fifth the separation between Mizar and Alcor in the Dipper.

This was something the astronomers couldn't accept without a struggle. They reasoned that there must be an unknown body out in space beyond the orbit of Uranus that was preventing it from moving freely in its course about the Sun.

Of those who applied themselves to a study of these irregularities, two men, John Couch Adams of England and Urbain Jean Jacques Leverrier of France, independently came up with correct solutions that told just where to find the unknown body in the sky. Adams made the earlier computations but, unfortunately, he sent his predictions to an astronomer who failed to act on them promptly. In the meantime, Leverrier completed his calculations and submitted them to Professor Galle of Berlin with the following note:

> Direct your telescope to a point on the ecliptic in the constellation of Aquarius, in longitude 326 degrees, and you will find within a degree of that place a new planet, looking like a star of about the ninth magnitude, and having a perceptible disc.

A half-hour search on the night of September 23, 1846, showed the new planet, later named Neptune, within two Moon-diameters of the predicted point. Happily, credit for this triumph of mathematical astronomy has been shared equally by both Adams and Leverrier.

Neptune is almost a twin brother of Uranus in size, but it is a billion miles farther from the Sun and its day is believed to be about 18 hours long. It creeps along its orbit at about 3½ miles per second and takes 165 of our years to make a circuit of the Sun. Neptune has two moons, one of them larger than our Moon; the other, discovered in 1949, has the longest month of any satellite, taking nearly a year to revolve around the planet.

Don't expect to see Leverrier's "perceptible disc" when you search for Neptune with your binoculars, though the planet is readily visible there as a star of 8th magnitude. You will find its predicted path among the stars charted for each year in the astronomical periodicals and in some almanacs. Hopefully, as a star and planet gazer, you will welcome the challenge of such a search and will know the elation that attends the finding of this planet which for more than eighty years was regarded as the outpost of the solar system.

Pluto

Even with the discovery of Neptune, all of the eccentricities in the erratic course of Uranus had not been explained. Accounting for these small perturbations required the presence of still another unknown body out in space and, not so long ago, another one was found.

The story of the discovery of Pluto is, in many ways, the telling of a twice-told tale for, once again, two men independently attacked the knotty problem of locating the unseen influential stranger. Once again, mathematics pointed the telescopes to the sky and, again, both targets were in close agreement. Beginning not long after the turn of the century, William H. Pickering in Jamaica and Percival Lowell in Flagstaff, Arizona, labored in this search for many years in an era when a computer was still a person equipped only with a pencil.

Not until 1930, when both men were dead, did success arrive. A comparison of two photographs taken of the same region on different nights showed a slight displacement in one faint starlike image. Clyde Tombaugh, an assistant at Lowell Observatory, photographically detected Pluto and announced its discovery on March 13, 1930, the anniversary of the discovery of Uranus and also the birthday of Percival Lowell, at whose observatory the discovery was made. As a further tribute, the first two letters of Pluto are also the initials of Percival Lowell, and these are combined in the symbol used for the newly found planet (♇).

Pluto is a small planet somewhat larger than Mercury with a density almost like that of Earth. From its distance of 3.5 billion miles, the Sun appears as no more than a brilliant starlike point of light; therefore, the temperature of this frigid little body is about -400 degrees Fahrenheit. Pluto completes a trip around the Sun in 248 of our years, traveling at a speed of only 3 miles per second. Its orbit is so eccentric that at perihelion, when it is nearest to the Sun, it is actually closer to the Sun than Neptune is. In fact, Pluto crossed Neptune's orbit in 1979 and will remain within it until the year 1999.

As recently as 1978 it was learned that little Pluto has a satellite circling around it at a distance of only 12,000 miles. Inasmuch as this satellite, since named Charon, is only two or three magnitudes fainter than Pluto itself, this may almost be a case of the tail wagging the dog.

At its closest approach to Earth, Pluto is no brighter than a 13th-magnitude star, so don't bother to get out your binoculars. Instead, get them ready for a visit to our moon.

CHAPTER 14
The Moon

With the exception of the Sun, we see the Moon more often than any other celestial object, so we need no chart to show us where to find it. For billions of years, perhaps even from the very beginning of the solar system, the Moon has been the Earth's nearest neighbor and constant companion. It is the only other world that man has ever visited, the only one whose surface dust holds footprints from the Earth.

But long before that giant step was made in 1969 we had learned many of our neighbor's secrets. In the years that followed Galileo's first sighting of those distant "seas" and mountains, the earthward facing visage of the Moon has been measured more carefully and more precisely than have many of the features of the Earth. Quite literally, we have scratched the surface of the Moon, but we still know nothing of its origin or of its early history.

The Moon has a diameter of 2,160 miles, about one-fourth that of the Earth. It is so large in relation to the Earth that if seen from Mars or Venus this Earth-Moon system would appear as a large planet and a small planet slowly circling about their common center of gravity as they revolve around the Sun. The Moon's orbit around the Earth is, like all planetary orbits, an ellipse rather than a true circle. Its mean, or average, distance from us is 238,000 miles, though our 7x binoculars shrink this span so that we see the Moon just as though we might view it with the naked eye at a distance of 34,000 miles.

Viewing the Moon is, quite frankly, an instance where higher power binoculars would be helpful, for its surface is broad enough and bright enough that the dark and narrow field of higher powers presents no problem. However, the Moon is by far the easiest of all celestial objects to observe and your binoculars, when steadily held, have ample powers to guide you on many an evening moon-walk to hundreds of sunlit craters, seas, and mountains through all the ever-changing phases of our neighbor world.

It seems that these "ever-changing phases" have always been a little difficult to understand. Poets and novelists have often taken some strange liberties with the Queen of Night in their moonlight rhapsodies, for they have seen her slender crescent rising in the east at sunset or watched the Half Moon high up in the sky at midnight. My diagram on page 126 is intended to show you just how each phase is produced, but if you, too, find it quite a problem to visualize how our Moon, a round ball, can assume so many different shapes as seen from Earth in just one lunar month, this following bit of nighttime simulation should make it all as clear as day.

First, carefully peel half an orange. This now becomes the Moon with its two contrasting hemispheres. The white half from which the peel was removed is the hemisphere that is always in full sunlight; the darker half with the peel still on is the night side of the Moon, which is always in shadow. Now push a pencil or a knitting needle directly into the orange at the line where the two halves meet. This makes a convenient handle.

Near one corner of a darkened room place a single unshaded lamp to represent the Sun. Standing near the opposite corner of the room, hold the orange by its upright pencil axis directly between the light and your head, which becomes the Earth. The orange is now in the New Moon phase and quite invisible. Holding the orange rigidly before you with the peeled half facing you, slowly turn to your left. As you do so, the lighted portion of the orange will become visible to you, first as a slender crescent representing the crescent Moon as seen in the western evening sky. When you have made a quarter turn and face directly to your left, one-half of the peeled portion of the orange will now be visible, just as in the Half-Moon or First Quarter phase. As you continue turning to your left, the white portion first assumes a football, or gibbous, phase which gets larger

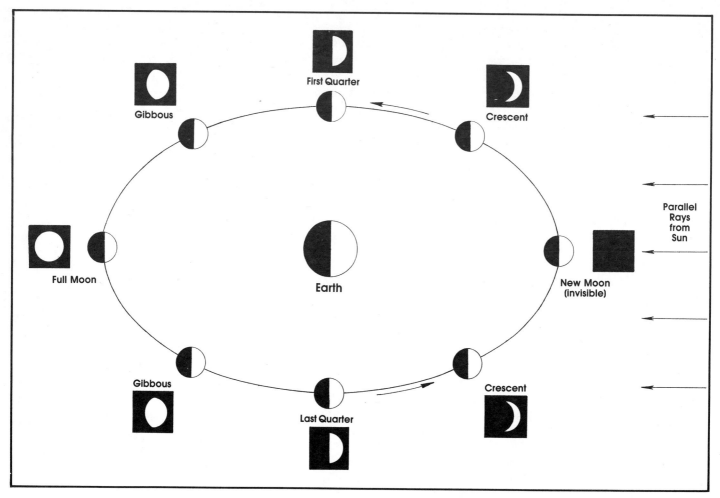

Phases of the Moon
The circling moons are shown as viewed from a point high above the orbit. The framed phases are shown as we see them from Earth.

THE MOON

until, when you have made a half turn, you now have the Full Moon phase. Turning still farther to your left, you'll come to each of these phases in reverse order until, after one complete turn, you'll come to the New Moon phase again.

As you performed this simple demonstration, you may have noticed some peculiar side effects that came to light in this cycle of the lunar phases. The first of these occurred in the crescent phase and is known as "the old moon in the new moon's arms." You may have noticed on seeing the crescent Moon in the west just after dark that you can see the entire ball of the Moon, though only the thin crescent is brightly lighted by the Sun. This is the effect of "earthshine" falling on the Moon in the same way that moonlight falls upon the Earth. What you are seeing is, of course, all sunlight, but in this case it is twice reflected — from Earth to Moon and back again to Earth. In this interchange of light, the Moon comes out ahead. Its dark rocks reflect the Sun's light poorly, while the greater surface of the Earth facing toward the Moon together with the high albedo, or reflective index, of the clouds that often cover much of the Earth cause the earthshine to be bright enough for you to recognize some of the lunar features on the Moon's night side with your binoculars.

As you reached the Full Moon phase in your simulation of the lunar orbit, your hand-held Moon must have entered the shadow of your head that the lamp cast on the wall. At this point, your Moon was in eclipse, just as you may have seen our real Moon eclipsed by the shadow of the Earth. However, if you want to keep this simulation true to nature, you must learn to tilt the orbit of your half-peeled Moon, for the plane in which the Moon revolves is tilted 5 degrees to the plane of the ecliptic in which the Earth revolves about the Sun. Were it not for this small tilt, we on Earth would witness an eclipse of the Moon every month at the time of Full Moon as well as an eclipse of the Sun at every New Moon phase. To make this correction gradually, raise your arm as you approach Full Moon so that the orange passes above the shadow of your head. Then slowly lower your arm so the orange passes just beneath the lamp at the New Moon phase.

Eclipse of the Sun

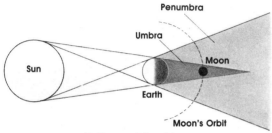

Eclipse of the Moon

Obviously, an eclipse of either the Sun or the Moon can only occur when Earth, Moon, and Sun are all nearly in line with each other. These line-ups can only take place at or near the nodes of the Moon's orbit. These are the two diametrically opposite points where the tilted orbit of the Moon crosses the ecliptic, the plane of the Earth's orbit around the Sun. At these crossings, the Moon must be in either its New or Full phase for an eclipse to occur.

Eclipses, both of Sun and Moon, take on added interest when viewed through binoculars. With the Sun, projecting its image with a telescope or a pinhole camera onto a piece of cardboard will show first contact quite precisely when the black body of the Moon takes its initial nibble from the Sun. Later you can watch as the growing ball of the Moon devours any sunspots that may be present on the earth-facing solar disk. At the last moment before totality, Baily's Beads — tiny flashes of sunlight shining through the valleys and mountain peaks of the Moon's eastern edge — will appear followed briefly by any red eruptive prominences that may be present around the Sun's rim.

The total eclipse of the Sun on February 26, 1979, was the last such performance to be seen in America during the present century. Total eclipses of the Moon, fortunately, are more obliging. Upcoming lunar blackouts have been scheduled for October 17, 1986, February 20, 1989, and August 17, 1989. Although they won't have the excitement, color, and scientific value of a solar eclipse, they will be visible to the entire portion of the Earth that faces the Moon at eclipse time — and has clear skies.

Because the Earth's shadow contains varying amounts of light refracted by our atmosphere, the Moon usually remains faintly visible during eclipse even when deep within the shadow of the Earth. The eclipse I watched in 1964 required some diligent searching before I was able to locate the Moon as a deep red ball with a slightly brighter rim. The Swedish astronomer Wargentin, for whom a crater has been named, tells of a total blackout that occurred on May 18, 1761: "The Moon's body disappeared so completely that not the slightest trace of any portion of the lunar disc could be discerned, either with the naked eye or with the telescope."

Exploring the Moon

When America was a New World, it was explored from east to west. The great explorers of that era first beheld the natural wonders of this country as they approached them from the east. Our exploration of the "New World" of the Moon will follow this, for we will proceed from east to west as we follow the advancing sunrise all the way across the rugged, pock-marked face of our neighbor world.

This jagged, ever-changing line that marks the division between night and day on the Moon is called the terminator. As you watch the phases of the Moon from New to Full then back to New again, you will see this slowly bending line first as a sunrise or morning terminator then, about two weeks later, it vanishes at the Full Moon phase. It then becomes a sunset or evening terminator that recedes night by night back to the New Moon phase once more.

Almost any calendar or almanac will give you the date, and often even the hour, when the New Moon phase occurs each month. The Moon at such times is, of course, quite invisible except on those occasions when we are blessed with an eclipse of the Sun and can watch the Moon, a round black ball, change from Old to New right before our eyes. Some clear spring evening, shortly after sunset, try to locate a 1-day-old Moon with your binoculars. With good seeing conditions and an open western horizon this shouldn't be too difficult, for such early sightings have been made when the Moon is only eighteen hours old. The Moon has even been seen just before sunrise as an Old Moon and then again as a New Moon just after sunset on the following evening. A 2-day-old Moon is usually easy to locate in the western evening sky; you can make out a few of the lunar features as the terminator is just beginning to reveal its rugged character.

As an aid in exploring the Moon, I have drawn a series of charts showing the Moon as seen with binoculars at four stages in its cycle from New to Full. They show the more prominent features as they appear 4, 7, 10, and 14 days after New Moon.

The 4-Day Moon

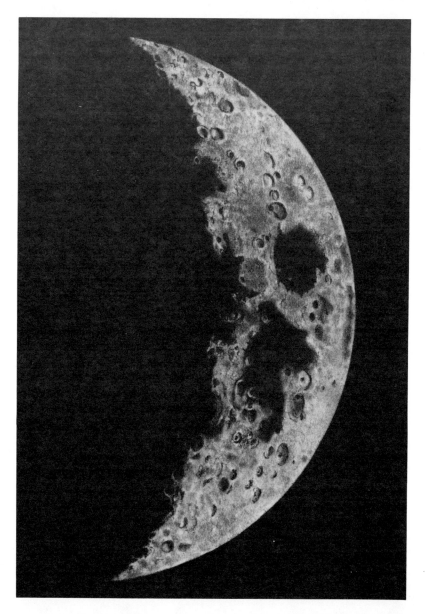

The 4-Day Moon

If you have had an opportunity to view the Moon each night since the New Moon, you have doubtless already noticed that the evenings of the second and third day of its cycle showed the "earthshine" phenomenon at its best, and you may have used your binoculars to grope your way about there in the dim light from Earth. On the 4-day Moon, the most outstanding feature is Mare Crisium, the Sea of Crises. This mare (pronounced MAH-re) can easily be seen near the Moon's upper right-hand edge. It is perhaps the first lunar object to which Galileo turned his new telescope when he began his study of the Moon in 1609. In binoculars, Mare Crisium appears like an isolated inland sea, so it's little wonder that many of the early observers thought of it and other similar dark-colored areas as being bodies of water. Nor is it surprising that when old constellation-maker Hevelius drew his map of the Moon he labeled each of these dark depressions a *mare*, a Latin word meaning sea.

In actual size, Crisium is 280 miles long by 360 miles wide — an area that would just contain the state of Missouri. Remember to look at Mare Crisium again a night or two after Full Moon. Sunset lighting will then be falling on its floor, which is the deepest of all the lunar maria, and there you may spot a couple of small craters on this ancient bed of once molten lava. Cape Agarum, a high promontory, can be seen jutting out into the mare from its southwestern shoreline.

South of Crisium, note Langrenus. It is usually described as a walled plain because of the extent of the enclosed area. Other similar though smaller features are often known as ringed plains and craters. Inasmuch as no one knows just what caused them — either volcanic action or meteoric impact — it is largely a matter of choice which term we use.

Still farther south of Langrenus and even larger in area is the walled plain Petavius. Note the mile-high mountain peak near the center of its floor. Just above Mare Crisium and touching its outer rampart is the oblong formation Cleomedes whose precipitous eastern wall rises two miles above the level of the floor. Farther north and close beside the present terminator are two craters lying side

The 7-Day Moon

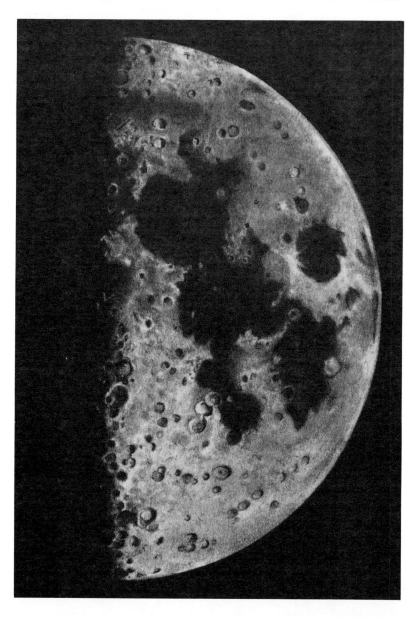

THE MOON

by side with their outer rims just touching. They immortalize the mighty heroes Atlas and Hercules, the latter being right on the sunrise line with its floor still deep in shadow. The large dark area south of Mare Crisium is Mare Foecunditatis, the Sea of Fecundity. Note Langrenus, already mentioned, on its eastern shoreline.

The 7-Day Moon

The chart of the 7-day Moon shows our satellite at its First Quarter or Half-Moon phase. Both of these terms are quite justified. The Moon has now completed the first quarter of its orbit around the Earth from New to New and at this point we can view just one-half of the Moon's surface that is visible from Earth. Note that many of the ring formations of the 4-day Moon, such as Langrenus and Cleomedes, that were then so clear and sharp, have lost much of their rugged character — along with their black outlining shadows — beneath the flatter lighting of the higher Sun.

A striking feature of the Half Moon is the dark expanse of four connected maria: Mare Serenitatis, Mare Tranquillitatis, Mare Foecunditatis and, to the south, Mare Nectaris, the Sea of Nectar. Look particularly at the group of three adjoining ring plains along the eastern shore of this sea of sweets. Your binoculars will clearly show all three. Theophilus to the north, with its prominent central peak and its high wall 64 miles in diameter, overlaps to the south the wall and plain of Cyrillus in a manner that clearly shows that Theophilus is the more recent formation. Note also the narrow pass or valley that connects Cyrillus and Catharina.

All along the now vertical terminator of the 7-day Moon you will find a maze of interesting craters and walled plains. Just below the center of the Moon look for another group of three walled plains that follow closely after Albategnius, the dark-floored formation that now shows as a deep crater right on the terminator. If you have the time, keep watching as the advancing sunlight begins to outline the high walls of these three plains. They are, from the north, Ptolemaeus, Alphonsus, and Arzachel. Do not confuse them with three similar craters that we found earlier on the shore of the Sea of Nectar.

The 10-Day Moon

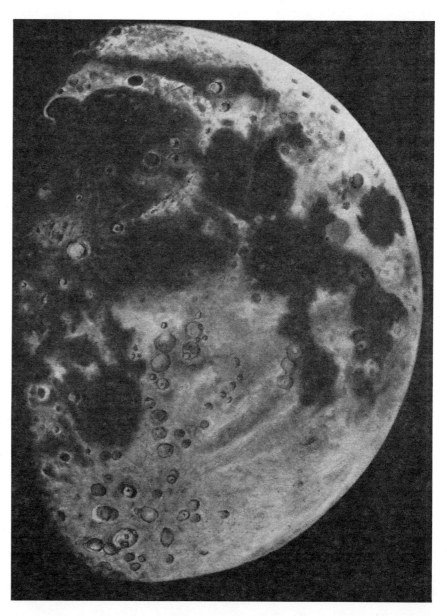

The 10-Day Moon

The Moon is now entering its gibbous phase as the terminator becomes a line of increasing convexity. Many of the Moon's most spectacular features are now coming into view. Your binoculars will reveal an assortment of craters, seas, and mountains that, were we closer to them, would prove as grand and awe-inspiring as anything our own Earth has to offer. Three of the Moon's most magnificent craters dominate the growing sector now lighted by the Sun. These are Plato (in the north), Copernicus (north of center), and Tycho (in the heart of a wild maze of ringed plains and craters to the south). All three are easily located with binoculars, and the fantastic terrain that surrounds each one will provide many an evening of visual exploration.

Plato lies on the northeastern shoreline of Mare Imbrium, the Sea of Showers. It is a walled plain rather than a crater, for its nearly circular, dark-colored floor is some 60 miles in diameter — large enough to contain the state of Delaware. It will appear quite level in your glasses and, unlike so many ringed formations we now have seen, it has no sign of any central peak or crater on its floor. To the west of Plato along this same northern shore the 10-day Moon now shows at its very best the Bay of Rainbows, Sinus Iridum. It is a broad, semi-circular bay with a cape at either end that seems to guard the entrance. This bay is one of the most beautiful and earthlike formations on the entire Moon. Note how the lighted northwestern wall of the bay juts into the darkness just beyond the terminator at this phase.

At the opposite side of this 700-mile-wide Sea of Showers, the shore is bordered by the Apennines, the Moon's finest mountain range. This chain is about 600 miles long and studded with peaks that rise nearly three miles above the level of the plain. This formation can be easily seen with the naked eye.

Within the curved sweep of the Apennines and the Caucasus Mountains, which form the western border of Mare Serenitatis, note the group of three walled plains: Archimedes, Aristillus, and Autolycus. North of this trio and forming the shore of Mare Imbrium from the Caucasus Mountains to the rim of Plato, you'll find the lunar

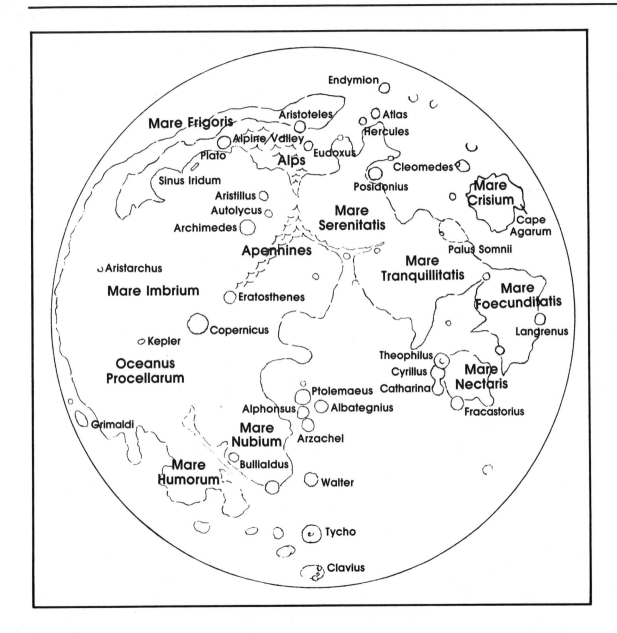

THE MOON

Best Viewing Times for Lunar Features

2 days	Mare Crisium, Langrenus, Vendelinus, Petavius, Cleomedes
3 days	Cape Agarum, Furnerius, Endymion
4 days	Mare Foecunditatis, Atlas, Hercules, Macrobius
5 days	Mare Nectaris, Palus Somnii, Piccolomini, Theophilus, Cyrillus, Catharina, Posidonius
6 days	Fracastorius, Maurolycus, Posidonius
7 days	Mare Tranquillitatis, Mare Serenitatis, Aristoteles, Eudoxus, Cassini, Hipparchus, Albategnius, Manilius
8 days	Alpine Valley, Alps, Appennines, Aristillus, Plato, Archimedes, Ptolemaeus, Alphonsus, Arzachel, Walter
9 days	Maginus, Tycho, Clavius, Eratosthenes
10 days	Mare Frigoris, Mare Imbrium, Longomontanus, Mare Nubium, Sinus Iridum, Copernicus, Bullialdus
11 days	Aristarchus, Kepler
12 days	Oceanus, Procellarum, Mare Humorum
13 days	Grimaldi
14 days	Lunar rays from Tycho, Kepler, Copernicus, Aristachus, Langrenus and Furnerius

Most of these formations are listed for the day on which they are on or near the terminator, though the maria are given for the time when the entire area of the mare can be seen.

Alps. Midway in this range look for the sharp crosswise slash of the 80-mile-long Alpine Valley. This unique feature is six miles wide with a level floor that ends among the the Alpine peaks. It has been suggested that this valley is a piece of land, bordered by two geological faults, that has collapsed.

The rugged Apennines terminate at the crater Eratosthenes and point onward to Copernicus, the finest ring-plain on the Moon and perhaps the best known lunar feature. This grand crater, first noted by Galileo, is 54 miles in diameter and three miles deep, with inner peaks and sharply terraced walls. In a telescope the entire plain surrounding Copernicus is thickly honeycombed with low hillocks and tiny craterlets.

Among the more conspicuous craters and ring-plains in the southern highland region is the ray-center Tycho, to which we will return when the Moon is Full. At the 10-day stage closely examine Clavius, an immense walled plain 140 miles across and the largest such feature on the visible lunar surface. There is a chain of smaller craters set in its broad floor.

Just before Full Moon, two contrasting craters can easily be located with your glasses. The first of these, Grimaldi, lies near the center of the Moon's western limb. This long oval is the darkest spot on the Moon. It is believed to be a walled plain that long ago was filled with lava, now solidified. Farther to the north, though still in this same immense Oceanus Procellarum, or Ocean of Storms, you can easily find the crater Aristarchus, which is the Moon's brightest spot. So dazzling is this crater that it can be made out even on the night side of the Moon. You can find it by earthshine alone when the Moon is still a slender sickle only two or three days old.

The Full Moon

When the Moon is Full the entire half that faces Earth is brightly lighted by the Sun, and therefore no shadows can be seen. We now view the Moon with the Sun almost directly behind us. Its rays fall perpendicularly on the lunar features and as a result many of the mountains and craters that we saw so clearly at earlier phases now have almost disappeared. The terminator — that advancing line of light and shadow that has led our searching glasses all the way across the Moon — has deserted us completely.

Despite this full-orbed lighting by the Sun, some aspects of our Moon are better observed at this stage than at any other. One of the improved features is the vast system of lunar rays that flares out in all directions, particularly from the crater Tycho, to give the Full Moon its often noted "peeled orange" appearance. Some of these rays extend straight across the surface for a thousand miles or more. One in particular, which originates on the rim of Tycho, flares northeastward across other craters and mountains lying along its course until its gray line completely bisects the Sea of Serenity, where you can follow its straight and narrow path from shore to shore with your binoculars. Other less extensive ray systems can be seen about the craters Copernicus, Aristarchus, and Kepler.

If clear skies have permitted you to follow the ever-changing terminator of the Moon in its nightly advance from east to west, you must have often marveled at how quickly the appearance of a mountain peak, a crater floor, or the shoreline of a sea can be altered by their shrinking shadows beneath the rising Sun. For each formation on the Moon there is a certain day between the New Moon and the Full Moon phases when it is at its very best. The table on page 138 gives the Moon's age — in even days after New Moon — when many of the objects shown on my charts may be observed to best advantage.

CHAPTER 15
Comets

There is a fascination about comets and their strange adventures that must appeal to nearly everyone. These ghostlike passers in the night are so completely different in character and make-up from anything you may have encountered in all your wanderings about the sky that I strongly urge you to spare no pains and miss no opportunities to turn your binoculars on every filmy voyager that comes within your range. This may require some patience, for comets of even binocular brightness don't visit us as often as we would wish. Although a dozen or more comets are announced every year, most of them are captured on photographic plates and few reach single-digit brightness.

If a comet of 7th magnitude or brighter appears in some region of the sky, your binoculars will afford the best possible means of finding the invader. The wide field of your low-power glasses will give you the most comprehensive and inspiring views of the complete tail structure and the overall extent of the brighter comets that do occasionally grace our skies.

The chance that some night your glasses may flush a comet of your own from the spangled cover of a star field is indeed a forlorn prospect, though it sometimes has happened in the past and, conceivably, it could occur again. Today, however, in all the dark-sky regions that permit a long exposure, it is usually the photographic plate that catches the first gleam of an approaching comet.

It would seem, then, that the comet-hunter who is short on instrumental power and long on perseverence would have the best chance of making a rare find by prowling in those twilight zones where the plate, because of fogging, is quite helpless. Furthermore, if we can judge by some past instances, a diligent search — with shielded objectives — almost to the solar limb with the Sun just below the horizon or with its bright disk hidden behind some distant tree or building might sometimes prove productive.

Every comet for at least a portion of its long career is a captive of the Sun. It is solely the attractive power of the Sun that draws a comet from the depths of space near enough to Earth for us to see it. Every comet swings about the Sun as it makes its perihelion, its closest approach, and then retreats back into space again. Comets are at their brightest when nearest the Sun, where they shine by reflected sunlight and by the Sun's excitation of their gases; therefore, it seems most logical to search for them in that region, even though their borrowed brilliance is greatly diminished by their bright sky background.

Comets approach the Sun from every possible direction, and the planes of their orbits lie at every conceivable angle to the plane of the ecliptic. It is quite possible for a comet to approach the Sun from the side opposite Earth and remain invisible to us until its swift motion suddenly brings it into view from behind the Sun. The first comet I ever saw did exactly that!

On January 18, 1910, astronomers at Lick Observatory in California first saw this comet in broad daylight just east of the Sun. They estimated its brightness as exceeding that of the planet Venus, which was then nearby in the sky. Several days and nights elapsed before I got to see it on the farm. In the meantime it had moved still farther to the east and by then I'm sure that it must have lost much of its brightness. It was the first comet of the year and so became known as Comet 1910a. I was only a youngster at the time and was not concerned with magnitudes and orbits, but I still vividly recall the ghostly appearance of this comet in our western evening sky. I remember well the sense of awe and wonderment that it aroused in me.

The brightest sun-grazing comet on record, and another daylight one, was that of 1843. In 1668, 1880, 1882, and 1887 bright comets were observed that wheeled about the Sun at some 300 miles per

Bennett's Comet in Morning Sky, April 3, 1970

second and were so near the solar surface that they must have passed right through its outer atmosphere. In more recent times, Comet Mrkos of 1957 and Comet Ikeya-Seki of 1965 were both sun-grazers, the latter being easily visible to the naked eye in full daylight by merely blocking out the sunlight with one's hand. Comets also have been found from time to time on those occasions when our Moon cooperated by cutting off the sunlight with a total solar eclipse. The first of these was recorded by the Roman statesman Seneca in the first century. Others occurred in A.D. 418, 1860, 1871, 1882, 1893, 1948, and more recently in 1970. This latter was first detected with binoculars when only 12 degrees from the Sun. An alert observer with shielded binoculars should detect any close Sun visitor more promptly and with more certainty than with the naked eye alone.

Binoculars are also useful in estimating the magnitudes of comets and conducting a night-to-night watch for any changes in their physical appearance or any sudden flare-ups in their brightness. Abrupt changes sometimes do take place in comets since they are rather frail affairs that are easily influenced by outside forces.

Estimating the brightness of a comet is a bit more complicated than judging the magnitude of a variable star. With the latter it is simply a matter of comparing the brightness of the variable with that of one or more stars of constant magnitude in the same field. However, since you won't find any convenient comparison comet to be pressed into service, you must make one from a star. To do this try to select a star of known brightness that, when thrown out of focus to match the apparent size of the comet will equal it in magnitude. The designated brightness of the star you use represents the total, or integrated, brightness of the comet.

If you dissect a typical comet you will find that it consists of three main parts: the nucleus, the coma, and the tail. The nucleus is the starlike point that contains the solid material of the comet in a spongy, porous aggregation of dust, solidified gases and ices of water, ammonia, methane, and carbon dioxide. The coma is the hazy envelope that surrounds the nucleus and is the one essential feature of every comet. It often is the only part that is visible. The coma is caused by the action of sunlight striking and driving off the gases and fine particles that compose the nucleus. A coma may be as big as a million miles in diameter.

The comet's tail is a continuation of the coma — the trailing out in space of the finer particles. It's nature's own bit of pollution, the "smoke" that is left behind as the comet is slowly consumed in the fire of the Sun's radiant energy. In this way the material that composes a comet's tail is forever lost to the comet, just as the smoke of any conflagration is dissipated in the air. It follows, then, that each time a comet passes near the Sun it leaves some of the substance of its tail behind and becomes a little smaller and less massive. This loss is seldom a perceptible one, however, for even though the tail may be 100 million miles in length, its density must be even less than that of the smoke to which I have compared it. Stars shine through a comet's tail with undiminished brightness.

As a general rule, a comet has no tail when first discovered in the mid-distant reaches of the solar system. At that point it appears as a hazy bit of light, not unlike a star seen through a mist. Only as the comet approaches the Sun, usually inside the orbit of the planet Mars, is the pressure of sunlight and the solar wind sufficiently great to repel the fine particles of the coma and form the tail. You can see this solar wind by its effect: Like a giant fan mounted on the Sun it always blows the comet's tail away from it, even as the comet sweeps around the Sun and, tail foremost, heads off into space.

Many astronomers believe that all comets move around the Sun in elliptical or closed orbits that after a period of time eventually bring them back to make another swing through perihelion. If this is correct, comets would be true members of the solar system, just as the planets are. Many of these periodic comets now are known and have been observed at successive returns. Most of these are classified as short-period comets. Perhaps the most frequently observed of these is Encke's Comet which has a period of only 3.3 years. This comet has often attained naked-eye brilliancy and should, when well placed in the sky, be an easy object for binoculars.

Among the slow-pokes of the skies are the long-period comets. This class includes many of the all-time greats: comets seen at midday close beside the Sun, awesome apparitions that creep across our skies but one time in a century or even in a thousand years. In this class, too, is the only comet that, if you are opportunely born and blessed with some longevity, you may see twice within your lifetime. This is Halley's Comet, the most famous of them all.

This historic comet has been seen at intervals of about 76 years ever since its first recorded appearance in 240 B.C. It is the first known periodic comet and the only one that has been bright enough at each return to be easily visible to the naked eye. Unlike most comets, this one was not named for its original discoverer — who may well have been some nameless cave man — but for the English scientist and Astronomer Royal, Edmond Halley, who discovered its periodicity and predicted its return. Halley watched this comet in 1682 and in plotting its orbit noticed in old records that similar bright comets had appeared in 1607 and 1531. As a period of about 76 years separated each of these three sightings, he suggested that the same comet was responsible. He then predicted that it would return in 1759; his forecast proved to be correct. Halley's Comet returned in 1835 and again in 1910, a sight that remains one of my own treasured childhood memories.

Halley's Comet has the longest period of any comet seen making more than one trip around the Sun. Its orbit is a long, slender ellipse that extends well beyond the orbit of Neptune at aphelion, its greatest distance from the Sun, and makes its perihelion passage between the orbits of Mercury and Venus. Thus it crosses the orbits of eight of the planets, though because of its inclined orbit it makes no close approach to any of them. Unlike the planets it moves around the Sun in retrograde motion — from east to west.

In 1948 Halley's Comet rounded the outermost loop of its orbit and in 1968 it recrossed Neptune's orbit. Slowly accelerating, it passed above the orbit of Uranus in 1979. By 1982, when between the orbits of Saturn and Jupiter, the comet was found at about magnitude 24 on photographs exposed for it, though it then showed but a fuzzy, starlike image devoid of any tail.

Backyard astronomers got their first look at Halley's Comet in October 1985 as it brightened to 8th magnitude. The comet appeared low on the horizon for observers in the Northern Hemisphere, who had to travel south to such places as Australia and South America to glimpse the comet when it was at its brightest between March and April 1986. But the disappointment of Halley's relatively poor performance was more than made up for by the first-ever views of a comet's nucleus provided by European and Soviet probes in March. From a distance of only a few hundred miles, the comet looked like a black, potato-shaped object spouting bright jets of dust and gas.

Let us hope that thoughtful parents of little children everywhere tried to make Halley's long-awaited visit an occasion and a memory that the child will treasure and remember throughout life. In doing this the parent has sponsored a potential future member of the world's oldest and most exclusive of all clubs, the Halley's Comet

Two-Timers Club, whose membership is restricted to those who have seen two visits of the comet. Obviously, to become a member you must be at least an octogenarian. The rarity of rarities in such a club would be a second-generation member. It's possible!

In the meantime, as we await the next coming of Halley's Comet and its hoped-for extravaganza, let's keep our eyes and glasses on the alert for the unexpected, which so often happens in the skies. Already, in this latter half of the twentieth century, we have met a number of interesting visitors, none of which had been predicted in advance. Early in the spring of 1957 the northwestern evening sky played host to naked-eye Comet Arend-Roland with its amazing spikelike tail that pointed *toward* the Sun! In August of that same year, the western skies were again enlivened, this time by 3rd-magnitude, tail-switching Comet Mrkos. In October 1965 Comet Ikeya-Seki cut its capers in broad daylight close beside the Sun before it followed its long glowing tail back into space.

While searching the southern skies from his home in South Africa, amateur John C. Bennett came upon a small, diffuse, tailless comet not far from the bright star Achernar in southern Eridanus. Though only 8th magnitude at the time of discovery, this comet soon showed signs of a brighter future as it moved northward toward its perihelion, which it reached on March 20, 1970. At that time it was estimated at zero magnitude, making the comet's total brightness nearly equal to that of Vega, then showing in the east. As I saw it in early April with its 15-degree-long divided tail spread out against the dark pre-dawn sky, I thought it easily the finest comet I had seen since 1910.

Scarcely had Comet Bennett slipped from the grasp of the largest telescopes when still another bright voyager was glowing in our eastern morning skies. This one, too, came from the Southern Hemisphere where, in November 1975, a 17th-magnitude comet was found on photographic plates exposed three months before at the European Southern Observatory in Chile by astronomer Richard West. Like its bright predecessor, Comet West was headed for its brief reunion with the Sun when found; it arrived at perihelion on February 25, 1976. As it swung around the Sun it was visible to the naked eye in daylight when it was estimated to be fully as bright as the planet Mars at a favorable opposition. The comet appeared at its best during the first week of March when it was at 1st magnitude against a darker pre-dawn sky. In my 7x35 binoculars, the tail was a streamered fan of marvelous complexity that curved to the north and west for some 25 degrees.

Comet West undoubtedly drew the largest audience of any comet in history, for it was well seen both north and south of the equator and all around the world. Its bright performance did much to atone for the dismal showing of stage-shy Comet Kohoutek three years before. As its finale, Comet West put on an act that even got an ovation from the astrophysicists: though it had come on stage with but a single head, it made its exit with no less than four! This quick-change act has been diagnosed as an early symptom of eventual disintegration, a penalty, perhaps, of a too-close approach during its command performance before the mighty Sun.

All who watched this splendid comet on those nights in early spring should treasure their remembrance of the occasion and rejoice at their good fortune to be in the audience of this one-time-only show. No Two-Timers Club will be formed to honor Comet West, for its period has been reckoned at around one million years.

CHAPTER 16
Meteors

During the hours you have spent getting acquainted with the stars and constellations, I'm sure that your searching eye was often attracted to the bright streak drawn on the dark ceiling of the sky by a shooting star or meteor. Literally the word meteor means "something of the air," and in the past the term has often been applied to clouds, rainbows, lightning, hail, and snow, as the poet Whittier did in his poem "Snow Bound":

> In starry flake and pellicle
> All day the hoary meteor fell.

Today such phenomena are recognized as the subjects of meteorology, the science of weather. Our present concern is with meteoritics, which treats the meteors of night, those swift visitors from the depths of space that, without warning, flash across our skies and disappear, leaving behind only a fading trail for us to try to trace back to its source and classify.

There are three distinct terms applied to these streaking voyagers, which I shall use here in the interests of accuracy, even though they all refer to the same object as it's seen in three different environments. First of all there is the term meteoroid, which applies to the object while it is still moving in outer space and invisible to us. If, however, this meteoroid enters the Earth's atmosphere, the resulting friction turns it into a glowing streak of light that now is called a meteor. In this stage it indeed is "something of the air." Finally, if the meteor survives the white heat of its passage through our atmosphere, it falls to Earth and becomes a meteorite, a suddenly

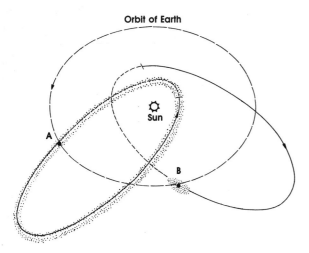

A: Earth passing through annual stream of meteoroids such as those of the Perseid shower.

B: Earth passing through swarm of meteoroids such as those of the Draconid shower.

The cross lines mark where the meteoroid orbits intersect the plane of Earth's orbit.

arrested visitor from outer space.

In the study of meteors, accurate tracking of their visible path across the sky is of great importance, for only in this way can their individual characteristics and family traits be learned. By estimating the brightness of these streaks, the size of a typical meteor has been found to be about that of a grain of sand or, at most, a small pebble. By simultaneously observing identical meteors from two stations a few miles apart, it has been determined, by triangulation, that meteors become visible when about 60 miles above us and cease to glow when about 40 miles high. By determining the height of meteors, the length of their trails, and the interval of their visibility, the average velocity of meteors has been found to be about 25 miles per second.

The hours from midnight to dawn are the best for meteor observing, though for the observer they also are often the coldest and most inconvenient. It's during these hours that the skywatcher is figuratively standing in the bow of the good ship Earth as it plows the seas of space. Thus he can clearly see all the flotsam that the ship encounters. On the other hand, in the period between dusk and midnight, the watch is from the ship's stern where the massive bulk of the ship itself blocks out everything but the slower movements of the ship's wake. My drawing on page 151 shows why this is so.

The Earth's two motions — its 18½-mile-per-second orbital movement around the Sun and its daily rotational turning on its axis — are both in the same direction, from west to east, and thus, in the after-midnight hours, the observer is moving into and meeting head-on any meteors in the Earth's path. This causes the after-midnight meteors to be swifter and brighter than those of the evening hours because the Earth's speed is added to the normal velocity of the meteor. The evening meteors are those that are catching up with the moving Earth, so the Earth's motion must be subtracted from that of the meteor, leaving it with a velocity of only about 7 miles per second.

Here is a case where haste makes waste. It is the slower-moving evening meteor that has the better chance of surviving the torrid passage through our atmosphere to land on Earth. Here, hopefully, it may be found and can spend its leisure years as a neatly labeled meteorite in some museum case where all can see and wonder at the awesome secrets it still withholds from us.

On any night of the year, if you are watching in a clear dark sky, an average of about six meteors per hour can be seen all about the sky moving in various directions. On rare occasions, perhaps a dozen times each year, you will be greeted by what appears to be an unusual number of meteors that seem to be streaking outward in all directions from a definite location in the sky. In this instance, you are watching a meteor shower. These showers occur on certain favored nights when the Earth cuts through the orbit of a meteor stream, which often may flow in a broad elliptical orbit all the way about the Sun. In my table on page 153, I've listed the most important of these showers together with the date of maximum activity and the number of meteors per hour that you may expect at an average occurrence.

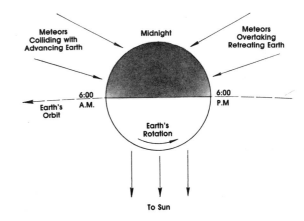

As an example, every year around the 12th of August the Earth moves through the orbit of a meteor stream. If the night is clear and there is no bright Moon in the sky, you are almost certain to see an unusual number of yellowish meteors streaking from the northeastern sky. If you face in that direction you may see as many as fifty of these per hour. This is the Perseid shower, so named because the streaks left by these meteors appear to intersect at a point in the constellation Perseus not far distant from the famous Double Cluster. From that point, these streaks seem to radiate outward in all directions like the spokes of a wheel from its hub. Actually these meteors are all moving parallel to each other, but because of the illusion of perspective, they seem to be coming from one point — called the radiant — just as the parallel rails of a railway seem to come together in the far distance.

The Perseids form a steady and dependable shower with about the same hourly rate every year. From this we know that the individual meteoroids are quite evenly distributed throughout the entire stream. For this reason, the Perseids are classified as an annual shower. In other showers with much more eccentric habits the meteoroids are mostly concentrated in one vast swarm somewhere along the stream's long course. These are the periodic showers; only at rare intervals does the Earth pass through one of these swarms and treat the fortunate observer to what must be the grandest of all sky shows. One such periodic shower, and the most noted of them all, is the Leonid shower, which on a number of occasions in the past has been seen to pour like a fiery fountain of stars from the Sickle in the constellation Leo.

The first Leonid shower to attract wide attention in America began on the evening of November 12, 1833. In the early morning hours at the height of the display it was estimated that as many as 35,000 meteors per hour were streaking across the sky in a shower that had now become a deluge. Here was clearly an instance where the Earth had plunged head-on into a dense swarm of meteoroids. Further investigation showed that this swarm was in an orbit that wheeled around the Sun and then extended out in space to a point somewhat beyond the orbit of the planet Uranus in a period of 33 years. Oddly enough this course proved to be an almost exact duplicate of the orbit of the Temple-Tuttle Comet of 1866, which also comes to perihelion every 33 years. Another brilliant meteor shower came in 1866, but 33 years later, in 1899, the expected bright rerun was a disappointment, for the meteor swarm in the meantime had passed too close to both Saturn and Jupiter and had been drawn slightly from its course. At the next time around, in 1932, the meteors came only at a rate of one per minute, and in most areas the Leonids were written off as a shooting star attraction.

But to the amazement and delight of all, in 1966 the meteor swarm was back on course again. In the western United States at maximum in the morning hours of November 17, meteors were falling far too fast for counting. The display was rated as equal to the all-time spectacular of 1833, with estimates of 60,000 to 150,000 meteors per hour. Unfortunately, in many eastern states on this occasion rain triumphed over the shower, and even where clear skies prevailed the peak of the display came after sunrise. The next return of the Leonid swarm is scheduled for November 1999, if all interfering planets will cooperate and keep their distance.

The fact that the orbit of the Leonid swarm is identical to that

Important Meteor Showers

Name of Shower	Date of Maximum	No. Per Hour	Parent Comet
Quadrantids	Jan. 1-3	40	Probably 1939-I
Lyrids	Apr. 21	15	1861-I
Eta Aquarids	May 4-6	20	Probably Halley
Delta Aquarids	July 29-30	20	?
Perseids	Aug. 12	50	1862-III
Draconids*	Oct. 9	100	Giacobini-Zinner
Orionids	Oct. 20	20	Halley
Taurids	Nov. 5	15	Encke
Leonids*	Nov. 16	15	1866-I
Geminids	Dec. 13	15	?

*Denotes a periodic shower. All others are annual showers.

Notes on Showers

Quadrantids — The radiant of this annual shower lies in the now obsolete asterism Quadrans Muralis in northern Bootes. As it's low in the sky in early January, this display is best seen in the morning hours of January 2.

Eta Aquarids — This shower leisurely drips from the Water Jar, which rises shortly before dawn. Best seen from southern stations.

Perseids — Occasional members of this faithful family may appear from mid-July until late August. These are swift, bright, streak-leaving meteors that are best watched after midnight when the radiant is high in the sky.

Draconids — Often called Giacobinids from its parent comet discovered in 1900. I watched the finest display of this periodic shower on the evening of October 9, 1946, when for a time more than 100 meteors per minute were streaking across the sky from their radiant in the Head of Draco. Some of these were slow-moving fireballs brighter than the planet Venus. For me, that shower was a sky-wide exhibition of celestial fireworks.

Leonids — A few Leonids may be seen every year in the early morning hours around mid-November. They increase considerably in the years immediately preceding and following the great showers of the 33-year periods. This indicates that some meteoroids are scattered throughout the entire stream and that the main swarm itself is somewhat elongated in form.

of a comet is not just a coincidence. It is, rather, a confirmation of the long-suspected close relationship between comets and meteors. Astronomers generally believe that meteoroid swarms are the debris of old comets that have partially or completely disintegrated. Most of the showers that we see have now been attributed to some known comet whose gradual deterioration has produced the stream of meteoroids that the Earth, in passing, may ignite into a fiery rain of shooting stars.

At the same time that you are watching the various shower meteors and tracing them to their radiant, you are certain to see other meteors that clearly do not belong to the shower in progress, for they don't share a common radiant. These are known as sporadic meteors: mavericks that have not, as yet, been observed as belonging to any known shower.

Binoculars have an important role in meteor observing, particularly in the recording of telescopic meteors, a term which, in a general way, refers to all meteors too faint to be caught with the naked eye. Here I shall designate as binocular meteors all those of 6th, 7th, and 8th magnitude, for these are well within the range of 7x glasses.

Although low-power binoculars easily bring you at least ten times the total number of fixed stars that you can see with the naked eye, this ratio doesn't hold true for meteors. The broad expanse of sky dome that your unaided eye sees streaked with 50 meteors per hour on an August night dwindles to a Dipper bowl of sky when seen in your binoculars. For many years, I recorded all the telescopic meteors I saw while comet-hunting and observing variable stars with a telescope that had a field of view of 1½ degrees. My best annual total was 160 meteors, of which 18 were seen on a single busy night.

In observing binocular meteors you should select an area that is well up in the sky in order to avoid atmospheric contamination. Since estimating magnitudes is important here, the field can be that of a variable star that has a good sequence of comparison stars between 5th and 9th magnitudes, such as the variable R Leonis. Tracings or photocopies of such a chart masked to the diameter of your binocular field of view can be used as charts on which to draw a pencil line to indicate the streak made by each meteor you see. The direction of motion is indicated by an arrow attached to the line, which is also numbered to identify the streak for any notes you may make concerning magnitude, color, velocity, and length of streak. For this last, you should know the diameter, in degrees, of your field of view.

If a considerable number of charted meteors are moving in the same direction it's quite possible that they are shower meteors, in which case you should try to locate the radiant, for this is the main objective in observing telescopic meteors. To do this, select a second field at right angles to the streaks of the first field and some distance away. If you get a second series of parallel lines here, their intersection with the first series should lead you to the radiant.

At the time of a meteor shower of considerable activity you'll find it interesting to locate and explore the precise radiant point with your binoculars. As you approach the radiant you'll note that the meteor trails become shorter and shorter until, at the exact center of the radiant, some show no lateral motion at all but appear as brightening starlike points, for now they are moving directly toward you.

Meteor observing provides an interesting occupation for your binoculars. It can be of scientific value, too, for there is much to learn and the field is uncrowded. If you feel the fascination of these sudden streaks across the skies and would like to delve still further into the secrets revealed in the last brief moments of these frail voyagers from outer space, apply for more complete information to:

> The American Meteor Society
> Department of Physics and Astronomy
> State University College
> Geneseo, NY 14454

CHAPTER 17
Endless Trails

It has been a long and winding trail that you have traveled since you found the Dipper standing upright on its handle in the northern winter sky. In the nights that followed you saw the stars and constellations of each succeeding season cross your skies to drop down in the west. Through darkened glasses you watched our own bright star, the Sun, and met many of the members of his planetary family. You explored the seas and craters of the Moon as you traced the slow advance of daybreak across its rough terrain. Further along, if good fortune favored you, your rambling sky path was briefly lighted by a comet's ghostly glow or by a stream of shooting stars.

In this year-long parade of celestial sights, you may have found some favorites you and your binoculars have returned to again and again because of some special interest which they aroused in you. It may well have been the vagaries of some variable star, an unpredicted drop in the light of R Coronae Borealis, or the transient glow of ancient Mira in a barren autumn field of Cetus. Or perhaps it was a group of sunspots that formed or faded right before your eyes as they crept across our day-star's filtered face.

Perhaps the challenge of some less-traveled trail has appealed to you with its positive assurance that right now out in space an undiscovered comet is headed toward its rendezvous with the Sun. But no matter what star-lighted bypath intrigues you and turns your binoculars skyward, it will be an endless trail you follow. Each path winds through fields of wonder and enchantment that will surely lure you on to new challenges and new fulfillments through all the seasons of the years to come.

Appendix I
Recommended Reading

Many beginners assume there exists somewhere the perfect book, computer program, or magazine that contains all the answers to every astronomical question they might have. "If I could only buy this one item," they think, "I'd be set for life." I wish this were true. The cold fact is, however, that no single source covers all of astronomy, and the more questions you have, the more books and periodicals you're going to have to look through.

The best friend any inquisitive amateur astronomer has is probably the public library. Find the largest collection you can get to conveniently and start learning their holdings inside out. It's almost impossible to overstate the importance of this. If you find yourself baffled by the library's card catalogue and the intricacies of the Dewey Decimal System (astronomy is catalogued in the 520s), don't hesitate to ask at the reference desk for help.

Bookstores that stock new astronomy books range from terrible to outstanding. Everything depends on how much local demand there is for the subject and how well the store's manager does his or her job. If you don't see what you want, ask. A bookstore can order any item that's in print — though you may have to wait a few weeks for it to arrive.

Another source to explore is second-hand bookshops, though finding one with a good selection of astronomy books is rare. Again, making your wants known to the owner will help.

So much for "information outlets." Now, what should you look for?

To get an overview of the science of astronomy as opposed to the practicalities of backyard observing, the best guide is any one of the many "astronomy for poets" textbooks. The market abounds in these, and new editions appear every year. Look for those with the most recent copyright dates.

For research results so new they haven't yet appeared in textbooks, regularly check magazines such as *Science News* (which appears weekly), *Astronomy, Sky & Telescope,* and *Scientific American* (all monthlies), and *Deep Sky* (a quarterly).

As you get more involved with astronomy, you may want to consider purchasing a telescope. An excellent place to start the process is by reading *The Light-Hearted Astronomer* by Ken Fulton (AstroMedia, 1984). Fulton's common sense advice is the perfect antidote to the shrill hype often found in many advertisements.

Fulton assumes that you are buying a telescope, but if you are interested in building one, get Richard Berry's *Build Your Own Telescope* (Scribners', 1985). It describes how to build five telescopes with no more than average handyman skills.

A good guide to what your new telescope (either bought or built) will show you is Terry Holt's *The Universe Next Door* (Scribners', 1985). Holt covers the essentials of telescope use and provides an up-to-date guide for observing the solar system.

Any backyard astronomer needs a star atlas and Wil Tirion's *Sky Atlas 2000.0* (Cambridge University Press, 1981) has become a standard fixture in almost every observer's library. And although it isn't essential, you can also get the two-volume catalogue that accompanies it, *Sky Catalogue 2000.0*, edited by Alan Hirschfeld and Roger W. Sinnott (Cambridge University Press, 1982 and 1985). The first volume contains basic data (names, positions, brightnesses, spectral types, distances, etc.) for some 45,000 stars down to magnitude 8.0. The second volume provides similar data for 21,000 double stars, variables, and non-stellar objects such as clusters, nebulae, and galaxies.

Rounding out the basic observing library is *Burnham's Celestial Handbook* (Dover, 1978). This three-volume work, compiled by Robert Burnham Jr., is a massive opus that provides summarized information, arranged by constellation, on a great many celestial objects beyond the solar system.

After you have outfitted yourself with these works, the bibliographic choices multiply greatly. Both the Holt and Berry books listed above have excellent bibliographies that can guide you from here.

— Robert Burnham
Astronomy Magazine

Appendix II
The Skies from a Swivel Chair

In a lifetime of watching the skies with a variety of instruments, I have found that nothing is more conducive to serious and consistent observing than the comfort and convenience of some form of observatory or observing station where all the necessary equipment is right at hand for immediate use. From an ordinary office swivel chair, I devised an observing chair that has proved to be ideal for use with hand-held binoculars. My drawing on page 162 shows how it was constructed.

So that the chair wouldn't be defaced in any way, a piece of wood about ¾" x 2" x 8" (A) with a straight outer edge was screwed to the underside of each of the chair arms (B). To each of these, a piece ¾" x 8" x 8" (C) was positioned with its top edge flush with the upper surface of the chair arm. This was fastened to (A) with two screws. Next two long pieces, about ¾" x 2" x 18" (D) were laid on the chair arms and firmly secured to the upper edge of (C) with screws. These provide a support for the ½" x 18" x 24" plywood desk (E) which is mounted just in front of the seated observer, where it pivots on a single carriage bolt which secures it to one desk support (D) as shown. This permits the desk to swing to one side for easy entrance and exit. This desk provides a solid surface for making notes or drawings, while the three cleated edges keep such equipment as charts, notebooks and flashlight from falling off.

Two long upright supports about ¾" x 2" x 20" (F) are bolted to the square boards (C) whose large surface area provides sufficient bearing friction to hold them in any position. The adjustable arm rests (G) are then bolted to the supports at any convenient angle.

In actual use, the observer sits comfortably in the chair with his head against the fully adjustable headrest, while his arms on the elevated armrests hold his binoculars steadily centered on any celestial object for as long as he desires — without discomfort to the muscles of his arms or neck.

The observer's feet must easily reach the ground in order to provide motion in both altitude (up and down) and azimuth (from side to side). Then, a push against the ground tilts the chair and observer in altitude from horizon to zenith, while sidewise stepping will produce a complete circle in azimuth. In this way, the observer can quickly point to any region of the sky.

Should the conventional casters of the chair prove too small to function well on uneven ground, they can be removed, and skids of wood or strap iron substituted and the entire chair moved from place to place to avoid trees, lights, or buildings. A tarpaulin or plastic sheet held down with a weight on each corner provides a waterproof cover.

This tiny open-air observatory has been a joy to use. Here the stargazer has all of his equipment right at hand before him. Here, with the aid of his red-filtered flashlight, he can consult a book or chart or make an entry in his notebook without a shift in his position. Here too, and most importantly, the ardent watcher can scan the skies throughout the year for, when warmly clad and booted, he can, in comfort, "walk" his way from star to star through all the sparkling nights of winter.

Appendix III
A Pronounciation Dictionary

Achernar (AY-kur-nar)
albedo (al-BEE-do)
Albireo (al-BEER-e-o)
Aldebaran (al-DEB-a-ran)
Alpha Centauri
   (AL-fa sen-TAW-ree)
Alphecca (al-FEK-a)
Alpheratz (al-FEE-ratz)
Andromeda (an-DROM-e-da)
Antares (an-TAIR-eez)
aphelion (af-FEE-lee-yon)
Aquarius (ah-KWAIR-ee-us)
Aquila (AK-wil-la)
Arcturus (arc-TEW-rus)
Aries (AIR-ih-eez)
Auriga (aw-RYE-ga)
Aurigae (aw-RYE-jee)

beta (BAY-ta)
Betelgeuse (BET-ul-jooz)
binary (BYE-na-ree)
Bootes (bo-OH-teez)

Camelopardalis
   (camel-o-PARD-a-lis)
Canes Venatici
   (KAY-neez ve-NAT-i-sye)
Canis Major (KAY-nis MAY-jer)
Canopus (ka-NO-pus)
Capella (ka-PELL-a)
Carina (ka-RYE-na)

Cassiopeia (kas-i-oh-PEE-ya)
Cepheid (SEFF-ee-id)
Cepheus (SEE-fee-yus)
Cetus (SEE-tus)
Chi Cygni (ky SIG-nye)
Coma Berenices
   (KOH-ma BARE-a-NYE-seez)
Cygnus (SIG-nus)

Deimos (DY-mose)
Delphinus (del-FYE-nus)
Deneb (DEN-eb)
Denebola (den-EB-o-la)
Draco (DRAY-ko)

Equuleus (e-KWOO-lee-us)
Eridanus (e-RID-a-nus)
eta (A-ta)
Europa (yoo-RO-pa)

Fabricius (fab-RISH-yus)
Fomalhaut (FOE-may-low)

galaxy (GAL-ax-ee)
Galileo (gal-a-LEE-o, gal-a-LAY-o)
Gemini (JEM-in-nye)

Halley (HAL-lee)
Hevelius (hev-EE-le-us)
Hyades (HY-a-deez)

Io (EYE-o)

Kochab (KO-kab)

Lacerta (la-SUR-ta)
Lepus (LEE-pus)
Leverrier (le-VARE-e-ay)
Libra (LYE-bra)
Lupus (LOO-pus)

Magellanic (ma-jel-LAN-ik)
mare (MAH-rey)
maria (mah-RE-a)
Messier (mess-ee-AY)
Mira (MYE-ra)
Mizar (MYE-zar)
Monoceros (mo-NOSS-er-os)

nebula (NEB-yoo-la)
nebulae (NEB-yoo-lee)
novae (NO-vee)

Ophiuchus (of-e-YOO-kus)
Orion (o-RYE-un)
Orionid (o-REY-on-id)

Pegasus (PEG-a-sus)
Perseus (PER-see-us)
Piazzi (pee-AHT-zee)
Pisces (PIS-eez)
Piscis Austrinus
   (PIS-is aw-STRY-nus)

Pleiades (PLEE-ya-deez)
Polaris (poe-LAIR-us)
Praesepe (pree-SEE-pee)
Procyon (PRO-see-on)
Ptolemy (TOL-e-mee)

Regulus (REG-yoo-lus)
Rigel (RYE-jell)
Roemer (RAY-mer)

Sagitta (sa-JIT-ta)
Sagittarius (saj-i-TAIR-ee-us)
Scutum (SKYOO-tum)
Sirius (SEER-ee-us)
Spica (SPY-ka)

Taurus (TAW-rus)
theta (THAY-ta)
Tycho Brahe (TY-ko BRA-hee)

Uranus (YOOR-a-nus,
   yoo-RA-nus)

Vega (VEE-ga)
Vulpecula (vul-PEK-yoo-la)

zodiac (ZOE-dee-ak)
zodiacal (zoe-DYE-a-kal)

Appendix IV
The Brightest Stars

The twenty-one stars listed below are in order of decreasing brightness. The stars marked with an asterisk cannot be seen from mid-northern latitudes, although Canopus can be briefly glimpsed in winter from our most southern states. Also listed is the constellation to which each star belongs as well as a space for you to record the date when you first identified the star.

Star	Constellation	Magnitude	Date
Sirius	Canis Major	−1.46	
*Canopus	Carina	−0.72	
Arcturus	Bootes	−0.04	
*Alpha Centauri	Centaurus	0.00	
Vega	Lyra	0.03	
Capella	Auriga	0.08	
Rigel	Orion	0.12	
Procyon	Canis Minor	0.38	
Betelgeuse	Orion	0.4 to 1.3	
*Achernar	Eridanus	0.46	
*Beta Centauri	Centaurus	0.61	
Altair	Aquila	0.77	
Aldebaran	Taurus	0.85	

Star	Constellation	Magnitude	Date
Antares	Scorpius	0.96	
Spica	Virgo	0.98	
Pollux	Gemini	1.14	
Fomalhaut	Piscis Austrinus	1.16	
Deneb	Cygnus	1.25	
Regulus	Leo	1.35	
*Alpha Crucis	Crux	1.41	
*Beta Crucis	Crux	1.88	

Appendix V
The Constellations

Constellation	Abbreviation	Date of First Sighting
Andromeda	And	
Antlia	Ant	
Apus	Aps	
Aquarius	Aqr	
Aquila	Aql	
Ara	Ara	
Aries	Ari	
Auriga	Aur	
Bootes	Boo	
Caelum	Cae	
Camelopardalis	Cam	
Cancer	Cnc	
Canes Venatici	CVn	
Canis Major	CMa	
Canis Minor	CMi	
Capricornus	Cap	
Carina	Car	
Cassiopeia	Cas	
Centaurus	Cen	

Constellation	Abbreviation	Date of First Sighting
Cepheus	Cep	
Cetus	Cet	
Chamaeleon	Cha	
Circinus	Cir	
Columba	Col	
Coma Berenices	Com	
Corona Australis	CrA	
Corona Borealis	CrB	
Corvus	Crv	
Crater	Crt	
Crux	Cru	
Cygnus	Cyg	
Delphinus	Del	
Dorado	Dor	
Draco	Dra	
Equuleus	Equ	
Eridanus	Eri	
Fornax	For	
Gemini	Gem	
Grus	Gru	
Hercules	Her	
Horologium	Hor	
Hydra	Hya	

Constellation	Abbreviation	Date of First Sighting
Hydrus	Hyi	
Indus	Ind	
Lacerta	Lac	
Leo	Leo	
Leo Minor	LMi	
Lepus	Lep	
Libra	Lib	
Lupus	Lup	
Lynx	Lyn	
Lyra	Lyr	
Mensa	Men	
Microscopium	Mic	
Monoceros	Mon	
Musca	Mus	
Norma	Nor	
Octans	Oct	
Ophiuchus	Oph	
Orion	Ori	
Pavo	Pav	
Pegasus	Peg	
Perseus	Per	
Phoenix	Phe	
Pictor	Pic	

THE CONSTELLATIONS

Constellation	Abbreviation	Date of First Sighting
Pisces	Psc	
Piscis Austrinus	PsA	
Puppis	Pup	
Pyxis	Pyx	
Reticulum	Ret	
Sagitta	Sge	
Sagittarius	Sgr	
Scorpius	Sco	
Sculptor	Scl	
Scutum	Sct	
Serpens	Ser	
Sextans	Sex	
Taurus	Tau	
Telescopium	Tel	
Triangulum	Tri	
Triangulum Australe	TrA	
Tucana	Tuc	
Ursa Major	UMa	
Ursa Minor	UMi	
Vela	Vel	
Virgo	Vir	
Volans	Vol	
Vulpecula	Vul	

Appendix VI
Life-List of Celestial Objects

Observed and recorded by _____

This list includes all the named features mentioned in the pages of this book that can be observed with low-power binoculars. Blank spaces have also been provided at the end of the list so that you can record names, dates, and comments relating to special events and sightings — such as eclipses, comets, meteors, novae, occultations, etc. — which cannot otherwise be classified here.

I urge you to record the date of each new object that you identify, together with any comments that you care to make. In this way your life-list becomes a personal diary of your hours with the stars and things celestial, a diary that will increase in interest and value with the passing years.

In order to save space, identification dates for first-magnitude stars and the constellations should be entered in the proper columns in Appendix IV and Appendix V. They aren't repeated here.

Object	Date	Comments
STARS		
Polaris		
The Pointers		
Alkaid		
Kochab		
Mintaka		
Orion's Belt		
Orion's Sword		
Zeta Tauri		
Pi Puppis		
Castor		
Thuban		

Object	Date	Comments
Gamma Cephei		
Alpha Cephei		
The Sickle		
Gamma Leonis		
Zeta Leonis		
Epsilon Leonis		
Denebola		
Tau Leonis		
Alphard		
Cor Caroli		
Alphecca		
Mu Cephei		
Beta Serpentis		
The Keystone		
Zeta Scorpii		
Beta Librae		
The Water Jar		
DOUBLE STARS		
Mizar		
Alcor		
Gamma Leporis		
Zeta Geminorum		
Delta Corvi		

Object	Date	Comments
Nu Draconis		
Delta Cephei		
Theta Serpentis		
Kappa Herculis		
Mu Scorpii		
Nu Scorpii		
Alpha Librae		
Epsilon Lyrae		
Zeta Lyrae		
Albireo		
61 Cygni		
Omicron Cygni		
Alpha Capricorni		
Beta Capricorni		
Omicron Capricorni		
Alpha Cassiopeia		
Beta Piscium		
66 Ceti		
Alpha Ceti		
NEBULAE AND CLUSTERS		
M-42 (The Orion Nebula)		
The Hyades		
M-1		

Object	Date	Comments
M-45 (The Pleiades)		
M-79		
M-41		
NGC-2244		
M-50		
M-55		
M-44 (The Beehive)		
M-67		
M-49		
M-58		
M-60		
M-87		
M-53		
M-64		
M-51 (Whirlpool Galaxy)		
M-3		
M-5		
M-13 (Hercules Cluster)		
M-92		
M-80		
M-4		
M-6		
M-7		

Object	Date	Comments
M-8 (Lagoon Nebula)		
M-20 (Trifid Nebula)		
M-17		
M-22		
M-24		
M-11		
M-71		
M-27 (Dumbbell Nebula)		
M-30		
M-52		
NGC-663		
Double Cluster		
M-34		
M-15		
M-2		
NGC-7009 (Saturn Nebula)		
M-77		
M-31 (Andromeda Galaxy)		
M-33		
M-36		
M-37		
M-38		

CELESTIAL OBJECTS

Object	Date	Comments
VARIABLE STARS		
Betelgeuse (Alpha Orionis)		
R Leporis		
Eta Geminorum		
Zeta Geminorum		
R Leonis		
R Hydrae		
R Coronae Borealis		
Mu Cephei		
Delta Cephei		
R Serpentis		
Alpha Herculis		
RY Sagittarii		
Beta Lyrae		
Chi Cygni		
Eta Aquilae		
R Scuti		
Gamma Cassiopeiae		
Algol (Beta Persei)		
Mira (Omicron Ceti)		

Object	Date	Comments
THE SUN		
Sunspots		
Umbra		
Penumbra		
Faculae		
THE PLANETS		
Mercury		
Transit		
Venus		
Phases		
Mars		
Retrograde motion		
Jupiter		
Io		
Europa		
Ganymede		
Callisto		
Saturn		
Titan		
Uranus		
Neptune		

Object	Date	Comments
THE MOON		
Earthshine		
Terminator		
1-day Moon		
Mare Crisium		
Mare Foecunditatis		
Mare Serenitatis		
Mare Tranquillitatis		
Mare Nectaris		
Mare Imbrium		
Mare Frigoris		
Mare Nubium		
Mare Humorum		
Langrenus		
Vendelinus		
Petavius		
Cleomedes		
Furnerius		
Atlas		
Hercules		
Endymion		
Macrobius		
Piccolomini		

Object	Date	Comments
Theophilus		
Cyrillus		
Catharina		
Posidonius		
Fracastorius		
Aristoteles		
Eudoxus		
Cassini		
Hipparchus		
Albategnius		
Manilius		
Aristillus		
Plato		
Archimedes		
Autolycus		
Ptolemaeus		
Alphonsus		
Arzachel		
Walter		
Maginus		
Tycho		
Clavius		
Eratosthenes		

Object	Date	Comments
Longomontanus		
Copernicus		
Bullialdus		
Aristarchus		
Kepler		
Grimaldi		
Oceanus Procellarum		
Cape Agarum		
Palus Somnii		
Alpine Valley		
The Alps		
The Apennines		
Sinus Iridium		
Lunar Rays		

Object	Date	Comments
ASTEROIDS		
Ceres		
Pallas		
Vesta		
METEOR SHOWERS		
Quadrantids		
Lyrids		
Eta Aquarids		
Delta Aquarids		
Perseids		
Draconids		
Orionids		
Taurids		
Leonids		
Geminids		

Index

American Association of Variable Star Observers, 89, 104
Andromeda (the Chained Maiden), 76
Antares, 58–59
Aquarius (the Water-Bearer), 74
Aquila (the Eagle), 62
Arcturus, 48–49
Aries (the Ram), 77
Asteroids, 113–114
Auriga (the Charioteer), 67, 70
Aurorae, 104
Baily's Beads. *See* Moon
Bayer, Johann, 92
Betelgeuse, 25
Big Dipper: the Pointers, 4; proper motion of, *diagram*, 5; magnitude of stars in, *diagram*, 11; seasonal movement of, 3–4, *diagram*, 2
Binary stars, 24
Binoculars, 13–18; Galilean field glasses, 14; how to choose, 17; magnification, 14–15; opera glasses, 13–14; prism, 14
Bootes (the Herdsman), 48–49
Camelopardalis (the Giraffe), 54
Cancer (the Crab), 43
Canis Major (the Greater Dog), 32
Canis Minor (the Little Dog), 33

Canis Venatici (the Hunting Dogs), 48
Capricornus (the Sea-Goat), 64
Cassiopeia (the Queen), 65–66; mythology of, 76
Celestial pole, 8. *See also* Pole stars
Cepheids, 90
Cepheus (the King), 54
Ceres, 113–114; discovery of, 77
Cetus (the Whale), 75–76
Chi Cygni, 92–93; *diagram*, 91
Columba (Noah's Dove), 31
Coma Berenices (Berenice's Hair), 47–48
Comets, 141–148; Bennett's, *illustration*, 143; composition of, 144; Halley's, 145–146; magnitudes of, 144; relation to meteors, 153–154; sun-grazers, 142–143; West, 147–148
Constellations, development of, 5–6; autumn, 65–78; spring, 36–49; summer, 50–64; winter, 21–35
Corona Borealis (the Northern Crown), 49
Corvus (the Crow), 46
Crater (the Cup), 46
Cygnus (the Swan), 61–62

Delphinus (the Dolphin), 63–64
Delta Cephei, 90–91; *diagram*, 87
Double stars, 21, 24
Draco (the Dragon), 51
Earthshine. *See* Moon
Eclipses: *diagram*, 128
Eclipsing binary, 66
Ecliptic, 106
Equuleus (the Little Horse), 64
Eridanus (the River), 31
Eta Aquilae, 91–92; *diagram*, 62
Fabricius, David, 92
Faculae. *See* Sunspots
Field glasses. *See* Binoculars
Galileo, 66, 115, 117
Gamma Cassiopeia, 91; *diagram*, 88
Gemini (the Twins), 33–34
Great Nebula of Orion, 28
Great Year, the. *See* Pole stars
Greek alphabet: *table*, 20
Halley, Edmond, 145–146
Halley's Comet. *See* Comets
Hercules (the Kneeler), 58
Herschel, William, 34
Hevelius, 54
Hipparchus, 95
Hyades, 29
Hydra (the Water Snake), 43, 46
Jupiter, 114–116; moons of, 115–116; observing, 114–115
Leo (the Lion), 42–43
Lepus (the Hare), 31
Libra (the Scales), 59
Little Dipper: brightest stars in, 10; description of, 8; as magnitude guide, *diagram*, 11
Long period stars, 92
Lowell, Percival: Martian observations, 111–112; search for Pluto, 123
Lyra (the Lyre), 60–61
Magnitude: definition of, 10; system, 10–11
Mare, 131
Mariner probes. *See* Mars
Mars, 110–113; Lowell's observations of, 111–112; Mariner probes of, 112–113; moons of, *diagram*, 113; oppositions of, *diagram*, 110; retrograde motion of, 111
Mercury, 106–107; transits of, 107
Meridian, 19
Messier objects, 30
Meteors, 149–155; definition of, 149; Leonid shower, 152; observing, 148, 154; Perseid shower, 151; relation to comets, 154; showers, 151, *table*, 153
Milky Way, the, 79–85; *illustration*, 83
Mira. *See* Omicron Ceti
Moon, 124–140; Baily's Beads, 128; earthshine on, 127; eclipses, 127–128; 4-day phase, 131–133; Full Moon phase, 139–140; phases of, *diagram*, 126; 7-day phase, 133–134; 10-day phase, 135, 139; terminator, 129
Mythology: Andromeda, 76; Coma Berenices, 47–48; Perseus, 76
Nebulae, 81
Neptune, 121–122; discovery of, 64, 121–122; observing, 122
Northern Cross. *See* Cygnus
Novae, 95–99; historic sightings, 95–96; hunting, 97–99; recurrent, 96–97; supernovae,

97; Tycho's Star, 66, 95, 97
Omicron Ceti, 92; *diagram*, 90
Open clusters: in the Hyades, 29; in the Pleiades, 30; in Puppis, 33
Opera glasses. *See* Binoculars
Ophiuchus (the Serpent-bearer), 55, 58
Orion (the Hunter), 25, 28–29
Pegasus (the Winged Horse), 71, 74
Penumbra. *See* Sunspots
Perseus (the Hero), 66–67; mythology of, 76
Pisces (the Fishes), 75
Piscis Austrinus (the Southern Fish), 71
Planets, 105–123; planetary configurations, *diagram*, 106
Pleiades, the, 30–31
Pluto, 122–123; discovery of, 34, 123
Polaris, 4, 8, 9
Pole stars, 36–37, 40; of the Great Year, 37, *diagram*, 40
Prism binoculars. *See* Binoculars
Ptolemy, 10
Pulsating stars. *See* Cepheids
Puppis (the Stern), 33
R Coronae Borealis, 93; *diagram*, 92
R Hydrae, 94; *diagram*, 94
R Leonis, 93; *diagram*, 92
R Leporis, 93; *diagram*, 91
R Scuti, 91; *diagram*, 62
RY Sagittarii, 93–94; *diagram*, 93
Red giants, 25
Sagitta (the Arrow), 63
Sagittarius (the Archer), 59–60
Saturn, 116–119; Galileo's observations, 117; ring system, 118
Scorpius (the Scorpion), 58–59
Scutum (the Shield), 62–63
Serpens (the Serpent), 55
Sextans (the Sextant), 46
Spiral galaxies, 82
Stars, 21, 24; apparent motion of, 4; Greek letter designations, 20; proper motion of, 5; red giants, 25. *See also* Variable stars
Sun, 100–104; eclipses of, *diagram*, 128; observing precautions, 101
Sunspots, 103–104; *illustration*, 101; faculae, 103; penumbra, 103; umbra, 103
Supernovae, 97
TX Draconis, 94; *diagram*, 94
Taurus (the Bull), 29–31
Triangulum (the Triangle), 77
Tycho Brahe, 95
Umbra. *See* Sunspots
Uranus, 119–121; discovery of, 34, 74, 119; Voyager 2 at, 120
Ursa Major: *diagram*, 6; stars in, 21
Ursa Minor. *See* Little Dipper
Variable stars, 86–94; estimating brightnesses, 88; first discovered, 75; naming systems, 87, 88–89; observing, 87–88; recording observations, 89
Venus, 107–109; observing, 108; phases of, 107; transits of, 108–109
Virgo (the Virgin), 46–47
Vulpecula (the Fox), 63
War of the Worlds, 112
White dwarfs, 32
Zenith, 20
Zodiac, 105

OBSERVER'S LOG

OBSERVER'S LOG

OBSERVER'S LOG

OBSERVER'S LOG

OBSERVER'S LOG

OBSERVER'S LOG

OBSERVER'S LOG